The Open University

M381 Number Theory and
Mathematical Logic

GW00598021

Number Theory **Unit 3**

Congruence

Prepared for the Course Team by Alan Best

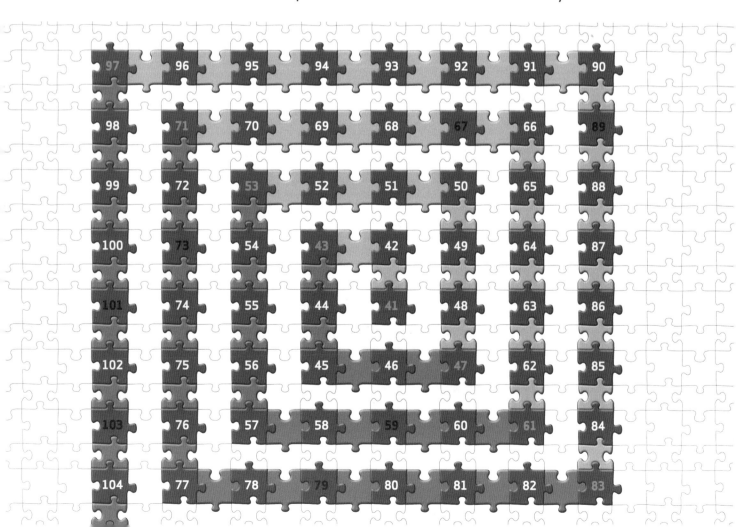

The M381 Number Theory Course Team

The Number Theory half of the course was produced by the following team:

Alan Best	*Author*
Andrew Brown	*Course Team Chair* and *Academic Editor*
Roberta Cheriyan	*Course Manager*
Bob Coates	*Critical Reader*
Dick Crabbe	*Publishing Editor*
Janis Gilbert	*Graphic Artist*
Derek Goldrei	*Critical Reader*
Caroline Husher	*Graphic Designer*
John Taylor	*Graphic Artist*

with valuable assistance from:

CMPU	*Mathematics and Computing, Course Materials Production Unit*
John Bayliss	*Reader*
Elizabeth Best	*Reader*
Jeremy Gray	*History Reader*
Alison Neil	*Reader*

The external assessor was:

Alex Wilkie	*Reader in Mathematical Logic, University of Oxford*

This publication forms part of an Open University course. Details of this and other Open University courses can be obtained from the Student Registration and Enquiry Service, The Open University, PO Box 197, Milton Keynes, MK7 6BJ, United Kingdom: tel. +44 (0)870 300 6090, e-mail general-enquiries@open.ac.uk

Alternatively, you may visit the Open University website at http://www.open.ac.uk where you can learn more about the wide range of courses and packs offered at all levels by The Open University.

To purchase a selection of Open University course materials, visit http://www.ouw.co.uk, or contact Open University Worldwide, Michael Young Building, Walton Hall, Milton Keynes, MK7 6AA, United Kingdom, for a brochure: tel. +44 (0)1908 858793, fax +44 (0)1908 858787, e-mail ouw-customer-services@open.ac.uk

The Open University, Walton Hall, Milton Keynes, MK7 6AA.

First published 1996. Reprinted 2001 and 2003. New edition 2007 with corrections.

Edited, designed and typeset by The Open University, using the Open University TeX System.

Printed and bound in the United Kingdom by The Charlesworth Group, Wakefield.

ISBN 978 0 7492 2269 7

2.1

CONTENTS

Introduction   4

1   **Properties of Congruence**   **6**
   1.1   Definition and basic properties   6
   1.2   Residue classes   9
   1.3   Further properties   12

2   **Divisibility Tests**   **14**

3   **Linear Congruences**   **16**
   3.1   Polynomial congruences   16
   3.2   Linear congruences   18
   3.3   Solving linear congruences   20

4   **Simultaneous Linear Congruences**   **22**
   4.1   The Chinese Remainder Theorem   22
   4.2   Solving simultaneous congruences   24

Additional Exercises   **27**

Solutions to the Problems   **30**

Solutions to Additional Exercises   **35**

Index   **43**

INTRODUCTION

It is curious how often what seem to be very different branches of mathematics become interlinked by some problem. For more than 2000 years it had been known how to construct both an equilateral triangle and a regular pentagon inscribed in a given circle using only compass and straightedge. But as the seventeenth century drew to its close constructions were still not known for any other regular polygon with a prime number of sides. The brilliant French mathematician Fermat had been discussing the numbers $F_n = 2^{2^n} + 1$ (later to become known as *Fermat Numbers*) believing them all to be prime, unaware of how intricately these numbers were woven in the solution of the polygon problem. It required the genius of the German mathematician Gauss to see how these numbers could be used to solve the construction problem.

Fermat will appear frequently in the remaining units of the course. We shall read more about him in *Unit 4*.

Carl Frederick Gauss (1777-1855)

Gauss was unquestionably one of the greatest mathematicians of all time. His father, a labourer in Brunswick, did not want his son to receive an education. But Carl was an infant prodigy, and his mother encouraged him in his studies. It is known that as a child of three he corrected an error his father made in a wages list, and by age ten he so amazed his schoolteacher with his arithmetic prowess that the teacher admitted there was no more he could teach the boy. The tale is reported that one day, to keep the children occupied, the teacher asked them to add up all the numbers from 1 to 100. Gauss handed in his slate immediately and to the dismay of the teacher was the only one in the class to get the correct answer. It is presumed that Gauss had discovered for himself the formula for the sum of the first n natural numbers.

Before he reached the age of twenty Gauss had, amongst many other achievements, conjectured the Prime Number Theorem, proved the Law of Quadratic Reciprocity (which we shall meet in *Unit 6*) and given a construction using compass and straightedge for a regular 17-gon to be inscribed in a given circle. It was this last result that launched Gauss on his career as a mathematician and it is the first result listed in a diary he kept of all his discoveries. The diary, which did not come to light until some forty years after his death, records brief statements of 146 results which amply display the genius of Gauss.

But it is his book 'Disquisitiones Arithmeticae', published when he was 24 years old, for which Gauss is best remembered. In this renowned work Gauss laid the foundations of modern Number Theory. It was here that he introduced and developed the language and notations of 'the algebra of congruence', and used this novel approach to produce elegant solutions to a variety of number-theoretic problems. Towards the end of the book Gauss returned to the construction of the regular 17-gon and generalized his earlier result to show exactly which regular N-gons can be constructed with compass and straightedge; the construction is possible if, and only if, $N = 2^r p_1 p_2 \ldots p_k$, where the primes p_i are distinct Fermat numbers and $r \geq 0$.

Sadly Gauss' career as a mathematician was foreshortened as his interest was caught by many other subjects. In particular a succession of discoveries gained him a deserved reputation in astronomy, and indeed for forty years from 1807 he was the director of the Göttingen Observatory. He is also well remembered for his work on magnetism, where a unit of magnetic intensity is the 'gauss', appropriately named after him. Late in life he expressed regret that he had abandoned mathematics, described by him as 'the queen of the sciences, with the theory of numbers being the queen of mathematics'. He always held number theory in the highest esteem and, not surprisingly, his name will feature regularly in the coming sections.

Construction of the regular pentagon. *AB* and *CD* are perpendicular diameters. *L* is the mid-point of *AO*. Using compasses: $LC = LM$; $CM = CN$. *CN* is the required edge.

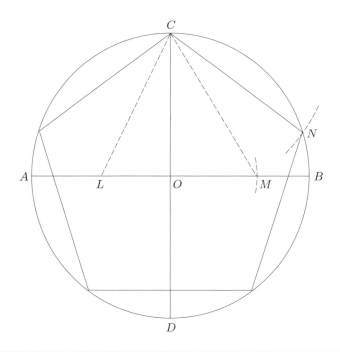

In the final section of the previous unit we introduced the idea of a remainder sequence, in which numbers were replaced by their remainder on dividing by some positive integer n. Observations about the behaviour of the remainder sequence were then projected back to reveal properties of the original number sequence. This was the underlying idea in Gauss' congruence. He developed an 'arithmetic of remainders', which he then put to work to ease computations in many areas.

In this unit we shall introduce Gauss' notion of congruence and develop properties of it which will be useful in problem solving throughout the remainder of the course. Here we shall restrict its application to a variety of problems involving remainders and, in particular, to solving linear congruences (which offer a more general way of looking at linear Diophantine equations). Finally, we shall prove the classical *Chinese Remainder Theorem*, which is concerned with the existence of (and how to find) solutions to a class of problems which can be modelled as a simultaneous system of linear congruences.

1 PROPERTIES OF CONGRUENCE

1.1 Definition and basic properties

It has been claimed that the result that p divides $2^{p-1} - 1$, for every odd prime p, was known to Chinese mathematicians over 2000 years ago. But how does one go about confirming instances of this result? If asked to show that $2^{16} - 1$ is divisible by 17 it would be feasible to compute 2^{16} by hand, subtract 1, and then carefully divide by 17. But it would be impractical to try the same line of attack to confirm that 641 divides $2^{640} - 1$ because this number is so huge. We shall see in this section that such divisions need not be as frightening as they first seem; we can investigate divisibility without having to compute the quotient involved in the division. We shall develop an 'arithmetic of remainders' in which everything hinges on one key definition:

Definition 1.1 Congruence modulo n

Let n be a fixed positive integer and let a and b be any integers. We say that a is *congruent modulo n to b* if the difference $a - b$ is divisible by n.

In symbols we write $a \equiv b \pmod{n}$.

If a is not congruent modulo n to b we say that a is *incongruent modulo n to b* and write $a \not\equiv b \pmod{n}$.

If we write $a \equiv b \pmod{n}$ it is to be understood that a and b are integers and n is a positive integer, for the statement would not make sense otherwise.

To illustrate the congruence notation consider the case of modulus $n = 11$. We have

$$3 \equiv 25 \pmod{11}, \text{ since } 3 - 25 = -22 \text{ is a multiple of } 11,$$

and

$$7 \equiv -26 \pmod{11}, \text{ since } 7 - (-26) = 33 = 3 \times 11.$$

On the other hand 12 is incongruent modulo 11 to 8, since $12 - 8 = 4$, and 4 is not a multiple of 11.

Problem 1.1 _____

For each of the following statements decide whether it is true or false.

(a) $3 \equiv 29 \pmod{8}$

(b) $3 \equiv -29 \pmod{8}$

(c) $63 \equiv 37 \pmod{13}$

(d) $37 \equiv 63 \pmod{13}$

(e) $63 \equiv 63 \pmod{37}$

(f) $63 \not\equiv 13 \pmod{25}$

Parts (c) and (d) of Problem 1.1 illustrate one apparent property of congruence, namely that if a is congruent modulo n to b then b is congruent modulo n to a. Part (e) illustrates the further property that any integer is congruent modulo n to itself. These are two of a number of general properties of congruence which highlight certain similarities with ordinary equality, and which we shall prove in our first result.

Theorem 1.1 *Properties of congruence*

Let n be a fixed positive integer and let a, b, c and d be any integers. Then the following properties hold.

(a) $a \equiv a \pmod{n}$.

(b) If $a \equiv b \pmod{n}$ then $b \equiv a \pmod{n}$.

(c) If $a \equiv b \pmod{n}$ and $b \equiv c \pmod{n}$ then $a \equiv c \pmod{n}$.

(d) If $a \equiv b \pmod{n}$ and $c \equiv d \pmod{n}$ then $a + c \equiv b + d \pmod{n}$.

(e) If $a \equiv b \pmod{n}$ and $c \equiv d \pmod{n}$ then $ac \equiv bd \pmod{n}$.

(f) If $a \equiv b \pmod{n}$ then $a^r \equiv b^r \pmod{n}$, for any integer $r \geq 1$.

Proof of Theorem 1.1

(a) $a - a = 0 = 0 \times n$, so $a \equiv a \pmod{n}$.

(b) If $a \equiv b \pmod{n}$ then there exists an integer k such that $a - b = kn$.

Then
$$b - a = -(a - b) = (-k)n,$$

so that $b \equiv a \pmod{n}$.

(c) If $a \equiv b \pmod{n}$ and $b \equiv c \pmod{n}$ then there exist integers s and t such that
$$a - b = sn \quad \text{and} \quad b - c = tn.$$

Adding these two equations:
$$(a - b) + (b - c) = sn + tn.$$

That is,
$$a - c = (s + t)n,$$

so that $a \equiv c \pmod{n}$.

(d) If $a \equiv b \pmod{n}$ and $c \equiv d \pmod{n}$ then there exist integers s and t such that
$$a - b = sn \quad \text{and} \quad c - d = tn.$$

Adding these two equations:
$$(a + c) - (b + d) = sn + tn = (s + t)n,$$

so that
$$a + c \equiv b + d \pmod{n}.$$

(e) With $a - b = sn$ and $c - d = tn$ as in the proof of property (d),
$$ac = (b + sn)(d + tn) = bd + n(bt + sd + stn),$$

and so
$$ac - bd = n(bt + sd + stn),$$

confirming that $ac \equiv bd \pmod{n}$.

(f) The case $r = 1$ is trivially true. Continuing by induction, suppose the result is true for $r = k$, that is, suppose $a^k \equiv b^k \pmod{n}$ whenever $a \equiv b \pmod{n}$.

Then
$$a^{k+1} = aa^k \equiv bb^k \pmod{n}, \quad \text{by Property (e)},$$
$$\equiv b^{k+1} \pmod{n}.$$

This confirms the truth for the case $r = k + 1$ and completes the induction step. ∎

The following problem gives you an opportunity to prove some simple properties of congruence working from the definition alone.

Problem 1.2 _____

Suppose that $a \equiv b \pmod{n}$ and m is an integer. Prove the following.

(a) If $m > 0$ and m divides n then $a \equiv b \pmod{m}$.

(b) If $m > 0$ then $ma \equiv mb \pmod{mn}$ and $ma \equiv mb \pmod{n}$.

(c) If $d > 0$ is a common divisor of a, b and n then $\dfrac{a}{d} \equiv \dfrac{b}{d} \left(\bmod \dfrac{n}{d}\right)$.

(d) $\gcd(a, n) = \gcd(b, n)$.

(e) For any integer c, $\quad a + c \equiv b + c \pmod{n}$.

Before developing further ideas, let us illustrate a range of problems in which properties of congruence can be exploited to ease computation.

Example 1.1

Show that $37^{37} + 2$ is divisible by 13.

We start by looking for numbers to which 37^{37} is congruent modulo 13. Now as $37 \equiv -2 \pmod{13}$, Theorem 1.1(f) tells us that

$$37^{37} \equiv (-2)^{37} \equiv (-1) \times 2^{37} \pmod{13}.$$

Now

$$2^4 = 16 \equiv 3 \pmod{13},$$

and so

$$37^{37} \equiv (-1) \times (2^4)^9 \times 2 \equiv (-1) \times 3^9 \times 2 \pmod{13}.$$

But

$$3^3 = 27 \equiv 1 \pmod{13},$$

so

$$37^{37} \equiv 3^3 \times 3^3 \times 3^3 \times (-2) \equiv -2 \pmod{13}.$$

Finally, Theorem 1.1(d) allows us to conclude that

$$37^{37} + 2 \equiv (-2) + 2 \equiv 0 \pmod{13},$$

which amounts to saying that $37^{37} + 2$ is divisible by 13. ◆

In the final step of Example 1.1 we inferred from a statement of the form $a \equiv 0 \pmod{n}$, that n divides a. This is an immediate consequence of the definition of congruence. Indeed we have a characterization of congruence modulo n in terms of the remainders on dividing by n, as given by the following.

Theorem 1.2 Congruent integers have the same remainders

$a \equiv b \pmod{n}$ if, and only if, the integers a and b have the same remainder when divided by n.

Proof of Theorem 1.2

Suppose first that a and b have the same remainder when divided by n, say

$$a = q_1 n + r \quad \text{and} \quad b = q_2 n + r, \quad \text{where } 0 \leq r < n.$$

Then $a - b = (q_1 - q_2)n$, showing that $a \equiv b \pmod{n}$.

Conversely, suppose that $a \equiv b \pmod{n}$, so that $a - b = kn$, for some integer k. Let b have remainder r when divided by n, so that $b = nq + r$ for some integer q. Then,

$$a = b + kn = nq + r + kn = (q + k)n + r,$$

shows that a has the same remainder, r, as does b. ∎

1.2 Residue classes

One outcome of Theorem 1.2 is that every integer is congruent modulo n to exactly one of the integers $0, 1, 2, \ldots, n - 1$, namely, whichever one of these is its remainder on dividing by n. This set of n integers is called the set of *least positive residues* modulo n. (Strictly speaking, since 0 is not a positive integer, it ought to be called the set of *least non-negative residues*, but this misnomer is firmly entrenched in the literature.)

Furthermore we can split up $\mathbb{Z}$ into non-overlapping sets called *residue* (or *congruence*) classes, where two integers belong to the same class if, and only if, they are congruent modulo n. For example, for the case $n = 4$ the residue classes are:

$$\{\ldots, -8, -4, 0, 4, 8, \ldots\} \ .$$
$$\{\ldots, -7, -3, 1, 5, 9, \ldots\}$$
$$\{\ldots, -6, -2, 2, 6, 10, \ldots\}$$
$$\{\ldots, -5, -1, 3, 7, 11, \ldots\}$$

A set obtained by selecting one member from each of these residue classes, for example $\{0, 1, 2, 3\}$ or $\{8, -3, 22, -9\}$, is called a *complete set of residues* modulo 4. In general, a set $\{a_1, a_2, a_3, \ldots, a_n\}$ of n integers is a complete set of residues modulo n provided no two of them are congruent modulo n. By virtue of Theorem 1.2 this will mean that the integers $\{a_1, a_2, a_3, \ldots, a_n\}$ are congruent modulo n, in some order, to $\{0, 1, 2, \ldots, n - 1\}$.

In choosing a complete set of residues modulo n, our most frequent selection will be the set of least positive residues, $\{0, 1, 2, \ldots, n - 1\}$. However we shall have occasion to use the set of *least absolute residues* obtained by choosing, from each residue class, one with least modulus. To be precise, from each residue class modulo n we choose the unique member x which lies in the range $-\frac{1}{2}n < x \leq \frac{1}{2}n$. For example, the set of least absolute residues for $n = 4$ is $\{-1, 0, 1, 2\}$ whilst for $n = 7$ it is $\{-3, -2, -1, 0, 1, 2, 3\}$.

Problem 1.3 ───────────────────────────────

Write down the set of least positive residues modulo n, and the set of least absolute residues modulo n, for (a) $n = 10$, and (b) $n = 11$.

Problem 1.4 ───────────────────────────────

Determine whether or not each of the following sets is a complete set of residues modulo 11.

(a) $\{6, 2, -3, 22, 23, -7, 7, 9, 43, -8, 16\}$;

(b) $\{0^2, 1^2, 2^2, 3^2, 4^2, 5^2, 6^2, 7^2, 8^2, 9^2, 10^2\}$;

(c) $\{0, 1, 2, 2^2, 2^3, 2^4, 2^5, 2^6, 2^7, 2^8, 2^9\}$.

Properties (d) and (e) of Theorem 1.1 are of great importance. In essence they say that the residue class to which a sum $a + c$, or a product ac, belongs does not depend so much on the individual numbers a and c but rather on the residue classes to which they belong. To qualify this remark, let us denote by $[r]$ the residue class modulo n to which the integer r belongs. So

$$[r] = \{\ldots, r - 2n, \ r - n, \ r, \ r + n, \ r + 2n, \ \ldots\}.$$

This will mean that residue classes are denoted in many different ways; for example, taking $n = 5$, $[-1] = [4] = [19]$ since $-1 \equiv 4 \equiv 19 \pmod 5$. In this notation properties (d) and (e) can be restated as follows.

(d)   If $[a] = [b]$ and $[c] = [d]$ then $[a + c] = [b + d]$.

(e)   If $[a] = [b]$ and $[c] = [d]$ then $[ac] = [bd]$.

These two facts amount to saying that the operations defined on the set of residue classes modulo n by

$$[a] + [c] = [a + c] \text{ and } [a] \times [c] = [ac]$$

are unambiguous. Whatever numbers are chosen to represent the two residue classes, the sum of the numbers will always belong to the same class, as will the product of the numbers.

Let

$$\mathbb{Z}_n = \{[0], \ [1], \ [2], \ldots, [n - 1]\}$$

be the set of all residue classes modulo n. Together with the operations of addition and multiplication defined above we have an arithmetic on $\mathbb{Z}_n$. For example, taking $n = 4$ we have the tables

+	[0]	[1]	[2]	[3]
[0]	[0]	[1]	[2]	[3]
[1]	[1]	[2]	[3]	[0]
[2]	[2]	[3]	[0]	[1]
[3]	[3]	[0]	[1]	[2]

×	[0]	[1]	[2]	[3]
[0]	[0]	[0]	[0]	[0]
[1]	[0]	[1]	[2]	[3]
[2]	[0]	[2]	[0]	[2]
[3]	[0]	[3]	[2]	[1]

where, for example, $[3] \times [2] = [2]$ is obtained from the definition

$$[3] \times [2] = [3 \times 2] = [6]$$

and the fact that $[6] = [2]$ since $6 \equiv 2 \pmod 4$.

Problem 1.5

Working in $\mathbb{Z}_{10}$, determine $[6] + [8]$ and $[6] \times [8]$ and find all $[b]$ such that $[6] \times [b] = [8]$.

Repeat the above, this time working in $\mathbb{Z}_{11}$.

Diversion

$$
\begin{array}{rcccl}
1 \times 8 & + & 1 & = & 9 \\
12 \times 8 & + & 2 & = & 98 \\
123 \times 8 & + & 3 & = & 987 \\
1234 \times 8 & + & 4 & = & 9876 \\
12345 \times 8 & + & 5 & = & 98765 \\
123456 \times 8 & + & 6 & = & 987654 \\
1234567 \times 8 & + & 7 & = & 9876543 \\
12345678 \times 8 & + & 8 & = & 98765432 \\
123456789 \times 8 & + & 9 & = & 987654321
\end{array}
$$

A problem in congruence modulo n is equivalent to a problem in $\mathbb{Z}_n$ with its arithmetic as illustrated in Problem 1.5 above. In practice, however, we shall abandon much of this formalism. What we shall do is select a complete set of residues, (usually the set of least positive residues), and work with these numbers, dropping the square brackets. When any sum or product is calculated we replace the answer by the chosen residue to which it is congruent. Here are a couple of illustrations of the way in which congruence is used.

Example 1.2

Show that $1^3 + 2^3 + 3^3 + \cdots + 60^3$ is divisible by 5.

In terms of congruence we are asked to show that $1^3 + 2^3 + 3^3 + \cdots + 60^3$ is congruent modulo 5 to 0. Now each of the terms r^3 is congruent modulo 5 to one of 0, 1, 2, 3 or 4, and so one line of attack would be to evaluate the least positive residue modulo 5 of each of these cubes and add up. However, a considerable short-cut is provided by Theorem 1.1(f), which tells us that if two numbers are congruent modulo 5 then their cubes are also congruent modulo 5. So, since $1 \equiv 6 \equiv 11 \equiv \cdots \equiv 56 \pmod 5$, we have

$$1^3 \equiv 6^3 \equiv 11^3 \equiv \cdots \equiv 56^3 \pmod 5.$$

Similarly

$$2^3 \equiv 7^3 \equiv 12^3 \equiv \cdots \equiv 57^3 \pmod 5;$$
$$3^3 \equiv 8^3 \equiv 13^3 \equiv \cdots \equiv 58^3 \pmod 5;$$
$$4^3 \equiv 9^3 \equiv 14^3 \equiv \cdots \equiv 59^3 \pmod 5;$$
$$5^3 \equiv 10^3 \equiv 15^3 \equiv \cdots \equiv 60^3 \pmod 5.$$

Therefore,

$$1^3 + 2^3 + 3^3 + \cdots + 60^3 \equiv 12 \times (1^3 + 2^3 + 3^3 + 4^3 + 5^3) \pmod 5.$$

Now,

$$2^3 \equiv 3, \ 3^3 = 27 \equiv 2, \ 4^3 = 64 \equiv 4 \text{ and } 5^3 \equiv 0^3 \equiv 0 \pmod 5,$$

and so

$$1^3 + 2^3 + 3^3 + \cdots + 60^3 \equiv 12 \times (1 + 3 + 2 + 4 + 0) = 120 \equiv 0 \pmod 5,$$

as claimed. ♦

There are simpler ways of solving Example 1.2. For instance, you may be familiar with the formula

$$\sum_{r=1}^{n} r^3 = \left(\frac{n(n+1)}{2} \right)^2$$

in which case, putting $n = 60$, the result drops out. Our solution here intentionally illustrates the use of the properties of congruence.

Example 1.3

Show that there do not exist integers x and y such that $x^2 + y^2 = 999$.

At first glance it might be difficult to see what this question has to do with congruence. In fact, it is a particular case of a general result concerning the possible values that a sum of two squares can take modulo 4.

Any integer x is congruent modulo 4 to one of the integers 0, 1, 2 or 3. Theorem 1.1(f) tells us that x^2 is therefore congruent modulo 4 to one of the numbers 0^2, 1^2, 2^2 or 3^2. That is, x^2 is congruent modulo 4 to either 0 or 1 (since $2^2 \equiv 0 \pmod 4$ and $3^2 \equiv 1 \pmod 4$). Similarly, for any integer y, y^2 is congruent modulo 4 to either 0 or 1. Theorem 1.1(d) now gives

$$x^2 + y^2 \equiv 0, 1 \text{ or } 2 \pmod 4$$

(as 0, 1 and 2 are the possible results of adding either 0 or 1 to either 0 or 1).

This argument tells us that any integer which is congruent modulo 4 to 3 cannot be written as the sum of two squares. Now $999 \equiv 3 \pmod 4$ is such an integer, and so the equation

$$x^2 + y^2 = 999$$

cannot have a solution in integers. ♦

And now some for you to try.

Problem 1.6 ─────────────────────────────

What is the remainder when $1! + 2! + 3! + \cdots + 1000!$ is divided by 7?

Problem 1.7 ─────────────────────────────

Let a be an integer which is not divisible by either 2 or 3. Prove that $a^2 - 1$ is divisible by 24. *Hint*: Show $a^2 \equiv 1 \pmod{24}$.

1.3 Further properties

There is a property of congruence which, although it is no more than a re-expression of an established divisibility property, will be used so often that it merits the status of a theorem.

Theorem 1.3

If $a \equiv b \pmod{m}$ and $a \equiv b \pmod{n}$, then

$$a \equiv b \pmod{\mathrm{lcm}(m, n)}.$$

Proof of Theorem 1.3

From the definition of congruence there exist integers r and s such that

$$a - b = rm \quad \text{and} \quad a - b = sn.$$

Let $d = \gcd(m, n)$, so that

$$m = dm' \quad \text{and} \quad n = dn', \quad \text{where } \gcd(m', n') = 1.$$

Theorem 4.5 of *Unit 1* establishes that $\gcd(m', n') = 1$.

Substituting for m and n in the equation $rm = sn$ gives

$$rdm' = sdn'; \text{ that is } rm' = sn'.$$

We infer from this that n' divides rm', whereupon Euclid's Lemma gives us that n' divides r. So putting $r = kn'$ we have

$$a - b = rm = kmn' = km\left(\frac{n}{d}\right) = k\left(\frac{mn}{d}\right) = k\,\mathrm{lcm}(m, n).$$

See Theorem 4.8 of *Unit 1*.

That is, $a \equiv b \pmod{\mathrm{lcm}(m, n)}$, as claimed. ∎

One particular case of Theorem 1.3 is well worth highlighting. We shall frequently want to apply this result when m and n are relatively prime, in which case $\mathrm{lcm}(m, n) = mn$.

Corollary to Theorem 1.3

If $a \equiv b \pmod{m}$ and $a \equiv b \pmod{n}$, where $\gcd(m, n) = 1$, then $a \equiv b \pmod{mn}$.

Notice how much simpler Problem 1.7 would have been with the benefit of Theorem 1.3. To show that $a^2 \equiv 1 \pmod{24}$ it is sufficient to show that $a^2 \equiv 1 \pmod{8}$ and $a^2 \equiv 1 \pmod{3}$. Both these separate parts are quickly confirmed from congruence ideas:

(1) a odd implies $a \equiv 1, 3, 5$ or $7 \pmod{8}$, and squaring,

$$a^2 \equiv 1 \pmod{8} \text{ in all cases;}$$

(2) a not divisible by 3 implies $a \equiv 1$ or $2 \pmod{3}$, and squaring,

$$a^2 \equiv 1 \pmod{3} \text{ in each case.}$$

We have seen that if $a \equiv b \pmod{n}$ then $ca \equiv cb \pmod{n}$, for any integer c. What about the converse of this result? If $ca \equiv cb \pmod{n}$ can we legitimately cancel the c's to recover $a \equiv b \pmod{n}$? In fact we cannot, as this simple counter-example shows:

$$2 \equiv 8 \pmod{6} \text{ but } 1 \not\equiv 4 \pmod{6}.$$

Here, cancellation of the common divisor 2 has produced a contradiction.

Common divisors on two sides of a congruence can always be cancelled, but in doing so the modulus has to be adjusted according to the following rule.

Theorem 1.4 Cancellation rule

If $ca \equiv cb \pmod{n}$ then $a \equiv b \left(\text{mod } \dfrac{n}{d}\right)$, where $d = \gcd(c, n)$.

Proof of Theorem 1.4

From $ca \equiv cb \pmod{n}$ we know that there exists an integer k such that $ca - cb = kn$. Now $\gcd(c, n) = d$, and so there exist relatively prime integers r and s such that $c = dr$ and $n = ds$. Therefore, substituting for c and n in the equation $ca - cb = kn$ gives

Theorem 4.5 of *Unit 1* established that r and s are relatively prime.

$$dra - drb = kds,$$

which, after cancelling the common d, gives

$$r(a - b) = ks.$$

From this s divides $r(a - b)$ and, as $\gcd(r, s) = 1$, Euclid's Lemma confirms that s divides $a - b$. Consequently $a \equiv b \pmod{s}$, where $s = \dfrac{n}{d}$. ∎

One special case of Theorem 1.4 is worth noting. When $\gcd(c, n) = 1$ the common divisor c can be cancelled without the need to adjust the modulus and is stated formally as follows.

Corollory to the cancellation rule

If $ca \equiv cb \pmod{n}$, where $\gcd(c, n) = 1$, then $a \equiv b \pmod{n}$.

In fact the two congruences in Theorem 1.4 are 'equivalent' because the reverse implication also holds. That is,

if $a \equiv b \left(\text{mod } \dfrac{n}{d}\right)$ then $ca \equiv cb \pmod{n}$, where $d = \gcd(c, n)$.

To justify this remark, suppose that $a \equiv b \left(\text{mod } \dfrac{n}{d}\right)$. Then there is an integer k such that

$$a - b = k\frac{n}{d}.$$

Multiplying by c:

$$ca - cb = \frac{kc}{d}n,$$

so that $ca \equiv cb \pmod{n}$, since $\dfrac{c}{d}$ is an integer.

We shall make use of the symbol $\Longleftrightarrow$ to indicate that two congruences are equivalent. For example, after cancellation of the common divisor 6 in the following,

$$12x \equiv 18 \pmod{27} \iff 2x \equiv 3 \pmod{9}.$$

Problem 1.8 _____

Use an appropriate cancellation to simplify each of the following congruences:

(a) $6x \equiv 18 \pmod{30}$;

(b) $20x \equiv 30 \pmod{45}$;

(c) $52x \equiv 39 \pmod{60}$.

Before we leave this area of congruence properties there is one final result which merits our attention. Euclid's Lemma for the case of a prime divisor p showed that if p divides ab, then either p divides a or p divides b. When this is expressed in congruence notation we have the result which follows. We shall have numerous occasions to refer to this result so, to ease identification of the reference, we shall continue to call it Euclid's Lemma for prime divisors.

Theorem 1.5 Euclid's Lemma for prime divisors

If $ab \equiv 0 \pmod{p}$ then $a \equiv 0 \pmod{p}$ or $b \equiv 0 \pmod{p}$.

In problem 1.8 we have made a start at solving congruence problems. We shall return to this in Section 3; but first we take a short detour.

2 DIVISIBILITY TESTS

Given any integer written in our familiar decimal representation, we can tell at a glance whether or not it is divisible by 2 or 5. To test divisibility by 5, for example, we simply look to see whether or not the units digit is 0 or 5. Though not quite so immediate, there are also fairly simple tests for divisibility by 3, 4, 8, 9 and 11. All these divisibility tests can be derived from congruence ideas.

Historically, the most significant divisibility test is that for division by 9.

> *A positive integer is divisible by 9 if, and only if, the sum of its digits is divisible by 9.*

In fact this is a special case of a stronger result which we shall prove shortly.

> *A positive integer is congruent modulo 9 to the sum of its digits.*

For example, if $N = 974\,064$ then

$$N \equiv 9 + 7 + 4 + 0 + 6 + 4 = 30 \equiv 3 \pmod{9}.$$

This fact led to a method of checking calculations, known as 'casting out nines' which, over the years, has been much used by book-keepers. The method amounts to reworking the calculations in the residue class arithmetic $\mathbb{Z}_9$. As an illustration, consider the multiplication shown in Figure 2.1. On the left the multiplication is carried out fully, whilst on the right the same calculation is carried out modulo 9.

$$
\begin{array}{r}
1\,4\,8\,9 \\
\times \quad 7\,5\,3 \\
\hline
1\,0\,4\,2\,3\,0\,0 \\
7\,4\,4\,5\,0 \\
4\,4\,6\,7 \\
\hline
1\,1\,2\,1\,2\,1\,7
\end{array}
$$

$$1 + 4 + 8 + 9 = 22 \equiv 4 \pmod{9}$$
$$7 + 5 + 3 = 15 \equiv 6 \pmod{9}$$

$$4 \times 6 = 24 \equiv 6 \pmod{9}$$

Check: $1 + 1 + 2 + 1 + 2 + 1 + 7 = 15 \equiv 6 \pmod{9}$

Figure 2.1   Casting out nines

The fact that the two answers agree modulo 9 is far from being conclusive proof that the calculation is correct. However, there is a good chance that a mistake in carrying out the calculation would have shown up during this check.

We shall prove the results giving divisibility tests for 9 and 11 together in the following theorem.

Theorem 2.1 Divisibility by 9 and 11

Let the integer N be written in decimal notation as
$N = a_m a_{m-1} \ldots a_2 a_1 a_0$ so that the digits satisfy $0 \le a_i \le 9$ with
$a_m \ne 0$. Then:

(a) $N \equiv a_0 + a_1 + a_2 + a_3 + \cdots + a_m \pmod 9$;

(b) $N \equiv a_0 - a_1 + a_2 - a_3 + \cdots + (-1)^m a_m \pmod{11}$.

The well-known test for division by 3: *A positive integer is divisible by 3 if, and only if, the sum of its digits is divisible by 3*, follows immediately from part (a).

Proof of Theorem 2.1

The key is to write out the number N in terms of powers of 10, namely

$$N = a_0 + 10 a_1 + 10^2 a_2 + 10^3 a_3 + \cdots + 10^m a_m.$$

(a) Since $10 \equiv 1 \pmod 9$, it follows from Theorem 1.1(f) that
$10^r \equiv 1 \pmod 9$ for all integers $r \ge 1$. Thus, from properties (d) and (e) of Theorem 1.1,

$$N \equiv a_0 + 1 a_1 + 1 a_2 + 1 a_3 + \cdots + 1 a_m \pmod 9$$
$$\equiv a_0 + a_1 + a_2 + a_3 + \cdots + a_m \pmod 9 .$$

(b) Similarly, since $10 \equiv -1 \pmod{11}$, $10^r \equiv (-1)^r \pmod{11}$ for all integers r and

$$N \equiv a_0 + (-1)a_1 + (-1)^2 a_2 + \cdots + (-1)^m a_m \pmod{11}$$
$$\equiv a_0 - a_1 + a_2 - a_3 + \cdots + (-1)^m a_m \pmod{11} . \qquad \blacksquare$$

In the congruences of parts (a) and (b) of Theorem 2.1, and in their proofs, we have unnecessarily used the digits of N in reverse order, starting at the units digit a_0. It will be helpful to use formula (b) in this way, starting at the positive units digit and then alternating the signs. But in using (a) we shall often, as we did in Figure 2.1 above, add the digits in their naturally occurring order.

Diversion

987654321×9	$=$	8888888889
987654321×18	$=$	17777777778
987654321×27	$=$	26666666667
987654321×36	$=$	35555555556
987654321×45	$=$	44444444445
987654321×54	$=$	53333333334
987654321×63	$=$	62222222223
987654321×72	$=$	71111111112
987654321×81	$=$	80000000001

Problem 2.1

Let $N = 9723$ and $M = 18\,056$.

(a) Determine the remainders when each of the numbers N, M, $N + M$ and $N - M$ are divided by (i) 9 and (ii) 11.

(b) Determine the missing digit (?) in the product

$$N \times M = 9723 \times 18\,056 = 175\,(?)58\,488$$

without using a calculator.

An integer N has an even number of digits. A second integer M is formed by moving the last digit of N to the front. (For example, if $N = 588\,471$ the 1 is moved to the front giving $M = 158\,847$.) Show that 9 divides $N - M$, 11 divides $N + M$ and 99 divides $N^2 - M^2$.

There are numerous variations on a trick which capitalizes on the connection between a number and the sum of its digits. Ask a person to give any 3-digit number (which is not palindromic). Without knowing what the number is, a prediction is made that application of the following sequence of operations on the number will produce the answer 1089.

Write down the number and the number formed by reversing the order of its digits. Then subtract the smaller of these numbers from the larger. Finally add to this the number formed by reversing its digits.

For example, if the chosen number is 638 then this is subtracted from 836 giving 198. Adding 198 to 891 gives 1089. The work of this section should enable you to explain why this is so.

It may happen that the first subtraction produces a 2-digit number. This is treated as a 3-digit number with leading 0 for the purposes of reversing its digits. For example, 99 would reverse to 990.

3 LINEAR CONGRUENCES

3.1 Polynomial congruences

Having studied basic properties of congruence, the next step is to look at the problem of solving congruences involving unknowns, such as $3x \equiv 7 \pmod{12}$ or $x^5 - x^3 \equiv 0 \pmod 8$. The first of these examples happens to have no solutions, for there is no integer x for which $3x \equiv 7 \pmod{12}$. On the other hand every integer x satisfies $x^5 - x^3 \equiv 0 \pmod 8$ — try one!

By an *integral polynomial* we mean a polynomial

$$P(x) = c_m x^m + c_{m-1} x^{m-1} + \cdots + c_1 x + c_0$$

in which the coefficients $c_0, c_1, \ldots, c_m$ and the variable x are integers. If $P(x)$ is an integral polynomial then the congruence $P(x) \equiv 0 \pmod n$ is a *polynomial congruence*. Our present interest is in solving polynomial congruences and we shall show straightaway that if an integer a is a solution of the congruence, that is, $P(a) \equiv 0 \pmod n$, then so too is every integer in the residue class modulo n of a.

Theorem 3.1

Let $P(x)$ be an integral polynomial. If $a \equiv b \pmod n$ then $P(a) \equiv P(b) \pmod n$. In particular, a is a solution of the polynomial congruence $P(x) \equiv 0 \pmod n$ if, and only if, b is a solution.

Proof of Theorem 3.1

Let $P(x) = c_m x^m + c_{m-1} x^{m-1} + \cdots + c_1 x + c_0$. Since $a \equiv b \pmod{n}$, properties (e) and (f) of Theorem 1.1 tell us that $c_r a^r \equiv c_r b^r \pmod{n}$, for each integer $r \geq 1$. Property (d) of Theorem 1.1 then gives

$$P(a) = c_m a^m + c_{m-1} a^{m-1} + \cdots + c_1 a + c_0$$
$$\equiv c_m b^m + c_{m-1} b^{m-1} + \cdots + c_1 b + c_0 \equiv P(b) \pmod{n}.$$

In particular, $P(a) \equiv 0 \pmod{n}$ if, and only if, $P(b) \equiv 0 \pmod{n}$. That is, a is a solution of the congruence $P(x) \equiv 0 \pmod{n}$ if, and only if, b is a solution. ∎

As a consequence of Theorem 3.1, it is apparent that the problem of solving the congruence $P(x) \equiv 0 \pmod{n}$ is really a problem in the residue class arithmetic $\mathbb{Z}_n$; if an integer is a solution then so too is every integer in its residue class. So the task is to find which residue classes solve the congruence. Two congruent integers which satisfy a polynomial congruence will be thought of as being the *same* solution.

Definition 3.1 Number of solutions of a polynomial congruence.

Let $S = \{b_1, b_2, \ldots, b_n\}$ be a complete set of residues modulo n. The *number of solutions* of the congruence $P(x) \equiv 0 \pmod{n}$ is the number of integers $b \in S$ for which $P(b) \equiv 0 \pmod{n}$.

Here is an example.

Example 3.1

Solve the polynomial congruence $x^2 - x \equiv 0 \pmod{6}$.

There are six residue classes modulo 6 and each may be a solution of the congruence. We choose the least positive residues, 0, 1, 2, 3, 4 and 5 and test each one in turn.

$$0^2 - 0 \equiv 0 \pmod{6}$$
$$1^2 - 1 \equiv 0 \pmod{6}$$
$$2^2 - 2 \equiv 2 \pmod{6}$$
$$3^2 - 3 \equiv 0 \pmod{6}$$
$$4^2 - 4 \equiv 0 \pmod{6}$$
$$5^2 - 5 \equiv 2 \pmod{6}$$

We see that 0, 1, 3 and 4 are solutions of the congruence and therefore so too is any integer congruent modulo 6 to one of 0, 1, 3 or 4. On the other hand any integer which is congruent modulo 6 to either 2 or 5 does not satisfy the congruence. The congruence has *four* solutions:

$$x \equiv 0, 1, 3, 4 \pmod{6}.$$ ◆

Problem 3.1

Solve the following polynomial congruences.

(a) $P_1(x) = x^2 + 3x + 4 \equiv 0 \pmod{7}$

(b) $P_2(x) = x^4 - 1 \equiv 0 \pmod{5}$

3.2 Linear congruences

We have quickly found an algorithmic method for solving polynomial congruences, namely, substitute for x, in turn, each value from any complete set of residues and record the resulting zeros. This is all well and good when the modulus in question is reasonably small and the degree of the polynomial is not too large. For more awkward polynomials the amount of computation involved in this sledgehammer attack quickly renders it impractical, and we must look around for something more subtle. In *Unit 4* we shall return to the problem of solving general polynomial congruences. There we shall discover ways of simplifying polynomial congruences and we shall be able to predict how many solutions a polynomial congruence will have. In *Unit 6* we shall be very much concerned with quadratic congruences and their solutions. For the remainder of this unit we shall confine our attention to the linear congruence

$$ax \equiv b \pmod{n}.$$

This method of solving a problem by checking every possible candidate is called *solution by exhaustion*.

This, the simplest of all polynomial congruences, crops up naturally in a variety of situations in number theory.

Suppose that x_0 is an integer such that $ax_0 \equiv b \pmod{n}$. This means that n divides $ax_0 - b$ or, put another way, there exists an integer y_0 such that

$$ax_0 - b = ny_0.$$

Rearranging, this equation tells us that $x = x_0$, $y = y_0$ is a solution of the linear Diophantine equation $ax - ny = b$. On the other hand if $x = x_0$, $y = y_0$ is any solution of $ax - ny = b$ then it is certainly the case that $ax_0 \equiv b \pmod{n}$. Thus the problem of finding all integers satisfying the linear congruence $ax \equiv b \pmod{n}$ is precisely the same problem as finding all x-values in solutions of the linear Diophantine equation $ax - ny = b$. As we have already solved the latter problem in Theorem 5.1 of *Unit 1*, our principal result on linear congruences is little more than a re-expression of that earlier theorem.

Theorem 3.2 Solution of linear congruences

Consider the linear congruence $ax \equiv b \pmod{n}$.

(a) The congruence has solutions if, and only if, $\gcd(a, n)$ divides b.

(b) If $\gcd(a, n) = 1$, the congruence has a unique solution.

(c) If $\gcd(a, n) = d$ and d divides b, then the congruence has d solutions which are given by the unique solution modulo $\dfrac{n}{d}$ of the congruence

$$\frac{a}{d}x \equiv \frac{b}{d} \left(\bmod \frac{n}{d}\right).$$

Proof of Theorem 3.2

(a) The linear Diophantine equation $ax - ny = b$ has solutions if and only if $\gcd(a, n)$ divides b. From the remarks preceding the theorem, it follows that the congruence $ax \equiv b \pmod{n}$ has solutions if, and only if, $\gcd(a, n)$ divides b.

(b) Suppose that $\gcd(a, n) = 1$. Theorem 5.1 of *Unit 1* tells us that if $x = x_0$, $y = y_0$ is one solution of $ax - ny = b$, then the general solution is

$$x = x_0 + nk, \quad y = y_0 + ak, \quad k \in \mathbb{Z}.$$

But for all integers k

$$x_0 + nk \equiv x_0 \pmod{n}$$

and so $x \equiv x_0 \pmod{n}$ is the only solution of $ax \equiv b \pmod{n}$.

(c) If $\gcd(a, n) = d$ and d divides b then

$$ax \equiv b \pmod{n} \iff \frac{a}{d}x \equiv \frac{b}{d} \left(\bmod \frac{n}{d}\right).$$

But $\gcd\left(\dfrac{a}{d}, \dfrac{n}{d}\right) = 1$, and so the right-hand congruence has a unique solution modulo $\dfrac{n}{d}$, say

$$x \equiv x_1 \left(\bmod \frac{n}{d}\right).$$

Hence the integers x which satisfy $ax \equiv b \pmod{n}$ are precisely those of the form $x = x_1 + k\dfrac{n}{d}$, for some integer k.

Consider the set of d integers

$$\left\{x_1, \; x_1 + \frac{n}{d}, \; x_1 + 2\frac{n}{d}, \dots, x_1 + (d-1)\frac{n}{d}\right\}.$$

Certainly no two of these integers are congruent modulo n because they are distinct and no two differ by as much as n. We claim further that, for any integer k, $x_1 + k\dfrac{n}{d}$ is congruent modulo n to one of them. To see why that is so, write $k = dq + r$, where $0 \le r < d$, as given by the Division Algorithm. Then

$$x_1 + k\frac{n}{d} = x_1 + (dq + r)\frac{n}{d} = x_1 + nq + r\frac{n}{d}$$

$$\equiv x_1 + r\frac{n}{d} \pmod{n}.$$

So these are the d solutions of $ax \equiv b \pmod{n}$. ∎

To illustrate part (c) and its proof, consider the congruence

$$12x \equiv 15 \pmod{21}.$$

Since $\gcd(12, 21) = 3$, and 3 divides 15, the congruence has three solutions modulo 21.

Cancellation of the common 3 reduces the congruence to

$$4x \equiv 5 \pmod{7}$$

which has a unique solution modulo 7, since $\gcd(4, 7) = 1$. That solution is

$$x \equiv 3 \pmod{7}$$

as can be seen from $4 \times 3 \equiv 5 \pmod{7}$. The argument in the proof of part (c) then supplies the three solutions to modulus 21 as

$$x \equiv 3 \pmod{21}, \; x \equiv 3 + 7 \equiv 10 \pmod{21} \text{ and } x \equiv 3 + 2 \times 7 \equiv 17 \pmod{21}.$$

Problem 3.2 _____

How many solutions does each of the following linear congruences have?

(a) $5x \equiv 13 \pmod{10}$

(b) $5x \equiv 10 \pmod{13}$

(c) $10x \equiv 5 \pmod{30}$

(d) $27x \equiv 318 \pmod{24}$

Diversion

Some squares containing the nine non-zero digits. There are 83 such squares in all.

$11826^2 = 139854276$	$25572^2 = 653927184$
$12543^2 = 157326849$	$15963^2 = 254817369$
$24237^2 = 587432169$	$29034^2 = 842973156$
$24441^2 = 597362481$	$30384^2 = 923187456$

3.3 Solving linear congruences

It is nice to have a theoretical result telling us that solutions exist but it is more important to be able to find those solutions.

Example 3.2

Solve the linear congruence $30x \equiv 15 \pmod{33}$.

As $\gcd(30, 33) = 3$ and 3 divides 15 Theorem 3.2 guarantees that the congruence has three solutions modulo 33. Failing all else we know that we could find these solutions by substituting for x each of the values $0, 1, 2, 3, \ldots, 32$ and recording which are solutions. But this last resort can usually be avoided, and can here, as we shall illustrate.

For a start we can use the cancellation laws to reduce the modulus. By the cancellation rule, Theorem 1.4, the common divisor 3 in both coefficients and modulus can be cancelled to give the equivalent congruence

$$10x \equiv 5 \pmod{11}.$$

The congruences are 'equivalent' in the sense that an integer x is a solution of one congruence if, and only if, it is a solution of the other.

As $\gcd(10, 11) = 1$ we know, by Theorem 3.2(b), that this congruence has a unique solution modulo 11.

Next we can cancel further. By the cancellation rule the common divisor 5 in the coefficients can be cancelled, this time the modulus remaining unchanged:

$$2x \equiv 1 \pmod{11}.$$

There are a number of ways of determining the unique solution modulo 11 from here. For example, we could replace the number 1 on the right-hand side of the congruence by 12, since $1 \equiv 12 \pmod{11}$:

$$2x \equiv 12 \pmod{11}.$$

Finally cancellation of the common divisor 2 of the coefficients yields the solution:

$$x \equiv 6 \pmod{11}.$$

Alternatively we could have multiplied through the congruence $2x \equiv 1 \pmod{11}$ by 6, obtaining

$$12x \equiv 6 \pmod{11},$$

whereupon replacement of the coefficient 12 by the congruent number 1 yields the same solution

$$x \equiv 6 \pmod{11}.$$

We noted that the original congruence had three solutions modulo 33 and we have ended with a unique solution modulo 11. There is no contradiction here, for in the set of least positive residues modulo 33 there are three numbers, namely 6, 17 and 28, which are congruent to 6 modulo 11. Consequently the unique solution $x \equiv 6 \pmod{11}$ is equivalent to the three solutions

$$x \equiv 6, 17, 28 \pmod{33},$$

for both determine the same set of integers, namely,

$$\{\ldots, -38, -27, -16, -5, 6, 17, 28, 39, 50, \ldots\}.$$

So either form can be taken as the answer here. Normally, except when the context of the problem demands otherwise, we shall always give the solution(s) in terms of the original modulus. ◆

Linear congruences, like the one we have just tackled, can all be solved using the following strategy.

Strategy: to solve the linear congruence $ax \equiv b \pmod{n}$

1   Check that $\gcd(a, n)$ divides b. If it does not the congruence has no solutions. If it does:

2   Cancel any common divisors of all three of a, b and n. The resulting congruence has a unique solution modulo the new modulus.

    The resulting coefficients (originally a and b) can then be changed by applying the remaining steps in any order, any number of times, with the goal of reaching a congruence in which the coefficient of x is 1.

3   Cancel any common divisor of the coefficients.

4   Replace either coefficient by any congruent number.

5   Multiply through the congruence by any number which is relatively prime to the modulus.

Step 5 needs a little clarification. To illustrate why the multiplier has to be relatively prime to the modulus, consider the congruence

$$7x \equiv 1 \pmod{40}.$$

If we multiply through by 6 we get

$$42x \equiv 6 \pmod{40},$$

and replacing 42 by the congruent number 2

$$2x \equiv 6 \pmod{40}.$$

Plainly $x \equiv 3 \pmod{40}$ is a solution of this congruence because $2 \times 3 \equiv 6 \pmod{40}$, but it is not a solution of the original congruence because $7 \times 3 = 21 \not\equiv 1 \pmod{40}$. So somewhere along the line an extraneous solution has been introduced.

The problem arose when we multiplied through the congruence by 6, because $\gcd(6, 40) > 1$. From $ax \equiv b \pmod{n}$ it is certainly true that $cax \equiv cb \pmod{n}$. But, according to Theorem 1.4, the cancellation rule, we can only reverse the argument and draw the conclusion $ax \equiv b \pmod{n}$ from $cax \equiv cb \pmod{n}$ when $\gcd(c, n) = 1$. A solution of $cax \equiv cb \pmod{n}$ is not necessarily a solution of $ax \equiv b \pmod{n}$ unless $\gcd(c, n) = 1$. Hence the relatively prime condition of Step 5 must hold before we multiply through a congruence.

The way to become proficient at solving linear congruences is through plenty of practice at applying the above strategy. We shall do one more example before inviting you to try your hand.

Example 3.3

Solve $9x \equiv 15 \pmod{26}$.

As $\gcd(9, 26) = 1$, Step 1 tells us that $9x \equiv 15 \pmod{26}$ has a unique solution modulo 26, and Step 2 is not applicable.

$$
\begin{aligned}
9x \equiv 15 \pmod{26} \iff& 3x \equiv 5 \pmod{26}, \quad \text{by Step 3,} \\
\iff& 27x \equiv 45 \pmod{26}, \quad \text{by Step 5, multiplying by 9,} \\
\iff& x \equiv 45 \pmod{26}, \quad \text{by Step 4,} \\
\iff& x \equiv 19 \pmod{26}, \quad \text{by Step 4.} \qquad \blacklozenge
\end{aligned}
$$

The steps that we followed in this example are certainly not unique. There are many ways of solving the congruence, but, of course, all must lead to the same conclusion.

Problem 3.3

Solve the following linear congruences.

(a) $28x \equiv 49 \pmod{53}$

(b) $34x \equiv 19 \pmod{52}$

(c) $137x \equiv 96 \pmod{19}$

(d) $51x \equiv 8 \pmod{53}$

(e) $18x \equiv 39 \pmod{69}$

4 SIMULTANEOUS LINEAR CONGRUENCES

4.1 The Chinese Remainder Theorem

In the first century AD the Chinese mathematician Sun-Tsu posed the following problem.

Example 4.1

Find a number which leaves remainders 2, 3 and 2 when divided by 3, 5 and 7 respectively.

In congruence notation the problem is asking for a simultaneous solution of the system

> By 'simultaneous solution' we mean an x which satisfies *all* the congruences.

$$x \equiv 2 \pmod 3; \quad x \equiv 3 \pmod 5; \quad x \equiv 2 \pmod 7.$$

We can solve this problem as follows. To satisfy the first congruence x must be of the form $3r + 2$. To satisfy the second congruence as well requires

$$3r + 2 \equiv 3 \pmod 5, \quad \text{i.e.} \quad 3r \equiv 1 \pmod 5.$$

This has the unique solution $r \equiv 2 \pmod 5$, i.e. $r = 5s + 2$ for some integer s. Therefore

$$x = 3r + 2 = 3(5s + 2) + 2 = 15s + 8.$$

This number has to satisfy the third and final congruence.

$$15s + 8 \equiv 2 \pmod 7 \iff 1s + 1 \equiv 2 \pmod 7$$

This has the unique solution $s \equiv 1 \pmod 7$. So $s = 7t + 1$ for some integer t and, substituting in the expression for x,

$$x = 15(7t + 1) + 8 = 105t + 23.$$

Any integer x of this form satisfies all three congruences. The least positive solution is 23 (corresponding to $t = 0$) and the complete solution is $x \equiv 23 \pmod{105}$. ♦

Consider now a general *system of simultaneous linear congruences* to be solved simultaneously, that is, a system

$$a_1 x \equiv b_1 \pmod{n_1}; \ a_2 x \equiv b_2 \pmod{n_2}; \ \ldots \ ; \ a_r x \equiv b_r \pmod{n_r}.$$

For an integer x to be a *simultaneous solution* of the system it must be a solution of each of the r individual congruences, and consequently a necessary condition for a solution is that

$$\gcd(a_i, n_i) \text{ divides } b_i, \quad \text{for each } i = 1, 2, \ldots, r.$$

Assuming that this condition holds, a first step in solving the simultaneous system is to solve each individual congruence, finding its unique solution to the appropriate modulus. From then on we can confine attention to solving simultaneous systems of the form

$$x \equiv b_1 \pmod{n_1}; \quad x \equiv b_2 \pmod{n_2}; \quad \ldots; \quad x \equiv b_r \pmod{n_r}$$

as in Sun Tsu's problem.

The fact that each congruence is solvable is not sufficient on its own for the simultaneous system to have a solution. For example, consider the following system.

$$x \equiv 1 \pmod{2}; \quad x \equiv 2 \pmod{4}$$

No even integer satisfies the first congruence and no odd integer satisfies the second; hence the simultaneous system has *no* solutions. Inconsistencies, such as the one illustrated here, can arise when two linear congruences have moduli which are not relatively prime. However, if we restrict ourselves to systems in which every pair of moduli are relatively prime, it turns out that simultaneous solutions always exist. In honour of their early contributions in this area, this result is known as the Chinese Remainder Theorem.

Theorem 4.1 Chinese Remainder Theorem

Let $n_1, n_2, \ldots, n_r$ be positive integers such that $\gcd(n_i, n_j) = 1$ for $i \neq j$. Then the system of linear congruences

$$x \equiv b_1 \pmod{n_1}; \quad x \equiv b_2 \pmod{n_2}; \quad \ldots; \quad x \equiv b_r \pmod{n_r}$$

has a simultaneous solution which is unique modulo $n_1 n_2 \ldots n_r$.

The moduli n_1, n_2, $\ldots$, n_r are said to be pairwise relatively prime.

Proof of Theorem 4.1

We shall prove the theorem by first constructing a solution and then showing that it is unique modulo $n_1 n_2 \ldots n_r$.

Let $N = n_1 n_2 \ldots n_r$ and $N_k = \dfrac{N}{n_k}$, for $k = 1, 2, \ldots, r$. Then, since $\gcd(n_i, n_k) = 1$ for each $i \neq k$, it follows that

$$\gcd(N_k, n_k) = \gcd(n_1 n_2 \ldots n_{k-1} n_{k+1} \ldots n_r, n_k) = 1,$$

and so the linear congruence $N_k x \equiv 1 \pmod{n_k}$ has a unique solution modulo n_k. Let this solution be x_k, so $N_k x_k \equiv 1 \pmod{n_k}$.

We shall show that the integer

$$x_0 = b_1 N_1 x_1 + b_2 N_2 x_2 + \cdots + b_r N_r x_r$$

satisfies each of the congruences. To achieve this we evaluate x_0 modulo n_k, for each $k = 1, 2, 3, \ldots, r$.

We start by noting that, whenever $i \neq k$, since n_k divides N_i, we have $N_i \equiv 0 \pmod{n_k}$. Consequently

$$x_0 \equiv b_k N_k x_k \pmod{n_k}, \quad \text{for each } k = 1, 2, \ldots, r,$$

as each of the remaining terms in the sum is congruent modulo n_k to 0.

Finally, since x_k was found such that $N_k x_k \equiv 1 \pmod{n_k}$, we have

$$x_0 \equiv b_k \pmod{n_k}$$

as claimed.

It remains to show that the discovered solution is unique modulo the product $n_1 n_2 \ldots n_r$. To this end suppose that x' is a second simultaneous solution of the system, that is,

$$x_0 \equiv x' \equiv b_k \pmod{n_k}, \quad \text{for each } k = 1, 2, \ldots, r.$$

This implies that each n_k divides $x' - x_0$, whereupon the fact that the moduli are pairwise relatively prime gives us that

$$n_1 n_2 \ldots n_r \text{ divides } x' - x_0.$$

This amounts to

$$x' \equiv x_0 \pmod{n_1 n_2 \ldots n_r}$$

exactly as required. ■

To illustrate the proof of the Chinese Remainder Theorem we trace its steps in an example.

Example 4.2

Solve the system of simultaneous linear congruences

$$x \equiv 5 \pmod 6; \quad x \equiv 4 \pmod{11}; \quad x \equiv 3 \pmod{17}.$$

The given values of the variables in the proof are:

$$\text{coefficients:} \quad b_1 = 5, \quad b_2 = 4, \quad b_3 = 3;$$
$$\text{moduli:} \quad n_1 = 6, \quad n_2 = 11, \quad n_3 = 17.$$

The calculated values are:

$$N = 6 \times 11 \times 17 = 1122; \ N_1 = 11 \times 17 = 187;$$
$$N_2 = 6 \times 17 = 102; \ N_3 = 6 \times 11 = 66.$$

The values x_i are obtained by solving the linear congruences $N_i x \equiv 1 \pmod{n_i}$.

$$
\begin{aligned}
N_1 x \equiv 1 \pmod{n_1} &\iff 187x \equiv 1 \pmod 6 \\
&\iff x \equiv 1 \pmod 6; \text{ so } x_1 = 1. \\
N_2 x \equiv 1 \pmod{n_2} &\iff 102x \equiv 1 \pmod{11} \\
&\iff 3x \equiv 1 \pmod{11} \\
&\iff x \equiv 4 \pmod{11}; \text{ so } x_2 = 4. \\
N_3 x \equiv 1 \pmod{n_3} &\iff 66x \equiv 1 \pmod{17} \\
&\iff -2x \equiv -16 \pmod{17} \\
&\iff x \equiv 8 \pmod{17}; \text{ so } x_3 = 8.
\end{aligned}
$$

Therefore

$$x_0 = 5 \times 187 \times 1 + 4 \times 102 \times 4 + 3 \times 66 \times 8 = 4151$$
$$\equiv 785 \pmod{1122}. \qquad \blacklozenge$$

Problem 4.1 —————————————————————————

Find the least positive integer n such that 3^2 divides $2n + 3$, 4^2 divides $3n + 4$ and 5^2 divides $4n + 5$.

4.2 Solving simultaneous congruences

One situation in which systems of simultaneous linear congruences can be introduced is when confronted with a problem of solving a single linear congruence where the modulus is large and composite. To see how this works let n have prime decomposition $p_1^{k_1} p_2^{k_2} \ldots p_r^{k_r}$ and compare the linear congruence

$$ax \equiv b \pmod n$$

and the system

$$ax \equiv b \left(\bmod\ p_1^{k_1}\right);\quad ax \equiv b \left(\bmod\ p_2^{k_2}\right);\quad \ldots\ ;\quad ax \equiv b \left(\bmod\ p_r^{k_r}\right).$$

Certainly any integer x which satisfies $ax \equiv b \pmod{n}$ also satisfies each congruence in the system because $p_i^{k_i}$ divides n for each i. On the other hand, since the p_i are pairwise relatively prime, repeated use of the corollary of Theorem 1.3 tells us that any solution of the simultaneous system must satisfy $ax \equiv b \pmod{n}$. So the two are equivalent.

We shall solve a problem of this sort in Example 4.3, which follows. However, in this example we shall abandon the formal approach of Example 4.2 and Problem 4.1 in which we traced the explicit construction of a solution as given in the proof of the Chinese Remainder Theorem. The method we shall adopt here, like our solution of Sun-Tsu's problem at the beginning of the section, will not require the same amount of careful arithmetic. Indeed, we often treat the Chinese Remainder Theorem as an important theoretical result which confirms the existence of solutions, but apply other methods, like the one we are about to illustrate, to discover what those solutions are.

Example 4.3

Solve the linear congruence $3x \equiv 5 \pmod{1001}$.

Observing that $1001 = 7 \times 11 \times 13$, we seek the simultaneous solution of

$$3x \equiv 5 \pmod{7};\quad 3x \equiv 5 \pmod{11};\quad 3x \equiv 5 \pmod{13}.$$

That is, (having first solved each of these congruences individually),

$$x \equiv 4 \pmod{7};\quad x \equiv 9 \pmod{11};\quad x \equiv 6 \pmod{13}.$$

The positive integers x which satisfy the congruence $x \equiv 6 \pmod{13}$ are

$$x = 6, 19, 32, 45, 58, 71, 84, 97, \ldots.$$

Now 97 is the first of these integers which is congruent modulo 11 to 9, and consequently, $x \equiv 97 \pmod{11 \times 13}$ is the unique solution of the final pair of congruences of the system. The positive integers in this solution set are

$$x = 97, 240, 383, 526, 669, \ldots.$$

Now 669 is the first of these which is also congruent modulo 7 to 4, and hence is the smallest positive integer satisfying all three congruences. Hence

$$x \equiv 669 \pmod{7 \times 11 \times 13}$$

is the unique solution of the congruence.

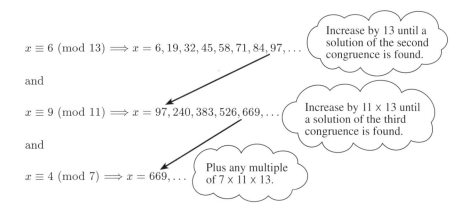

Figure 4.1   Solving simultaneous linear congruences   ♦

Problem 4.2 _____

Solve the linear congruence $7x \equiv 9 \pmod{165}$.

_____

There remains one loose end to tidy up. We have seen that when a system of linear congruences has moduli which are not pairwise relatively prime the system may be inconsistent and have no solutions. But this is not necessarily the case. For instance, the system

$$x \equiv 2 \pmod 6; \quad x \equiv 5 \pmod 9$$

involves moduli which are not pairwise relatively prime. However, $x = 14$ certainly satisfies both congruences, so solutions do exist.

Theorem 4.2

The system of linear congruences

$$x \equiv b_1 \pmod{n_1}; \ x \equiv b_2 \pmod{n_2}$$

has a simultaneous solution if, and only if, $\gcd(n_1, n_2)$ divides $b_2 - b_1$. If a solution exists then it is unique modulo $\mathrm{lcm}(n_1, n_2)$.

Proof of Theorem 4.2

Suppose that the integer x satisfies both

$$x \equiv b_1 \pmod{n_1} \quad \text{and} \quad x \equiv b_2 \pmod{n_2}.$$

That is, there exist integers r and s such that

$$x - b_1 = n_1 r \quad \text{and} \quad x - b_2 = n_2 s.$$

Eliminating x from this pair of equations leads to

$$b_2 - b_1 = n_1 r - n_2 s.$$

Now the right-hand side is an integer combination of n_1 and n_2 and, as such, is a multiple of $\gcd(n_1, n_2)$. Thus $\gcd(n_1, n_2)$ divides $b_2 - b_1$, and so this is a necessary condition for the system to have a solution.

To confirm that the condition is sufficient to guarantee a solution, we reverse the argument. Suppose that

$$b_2 - b_1 = k \gcd(n_1, n_2), \quad \text{for some integer } k.$$

Now we know that there exist integers u and v such that

$$\gcd(n_1, n_2) = u n_1 + v n_2.$$

Elimination of $\gcd(n_1, n_2)$ from this pair of equations produces

$$b_1 + k u n_1 = b_2 - k v n_2.$$

Then

$$b_1 + k u n_1 = b_1 + (ku) n_1 \equiv b_1 \pmod{n_1}$$

and

$$b_1 + k u n_1 = b_2 - (kv) n_2 \equiv b_2 \pmod{n_2}.$$

So $b_1 + k u n_1$ satisfies both congruences and confirms that simultaneous solutions do exist.

Finally, there is the question of uniqueness. Suppose that x_1 and x_2 are two solutions. That is

$$x_1 \equiv x_2 \equiv b_1 \pmod{n_1} \quad \text{and} \quad x_1 \equiv x_2 \equiv b_2 \pmod{n_2}.$$

Theorem 1.3 now completes the task:

$$x_1 \equiv x_2 \pmod{\mathrm{lcm}(n_1, n_2)}. \qquad \blacksquare$$

The result of Theorem 4.2 can be generalized to cover any number of congruences. A proof using mathematical induction is readily available but is omitted here as the details are somewhat tedious.

> ### Corollary
>
> The system of linear congruences
>
> $$x \equiv b_1 \;(\text{mod } n_1)\,; \quad x \equiv b_2 \;(\text{mod } n_2)\,; \quad \ldots\;; \quad x \equiv b_r \;(\text{mod } n_r)$$
>
> has a simultaneous solution if, and only if, $\gcd(n_i, n_j)$ divides $b_j - b_i$ for each pair i, j of suffixes. If a solution exists it is unique modulo $\text{lcm}(n_1, n_2, \ldots, n_r)$.

Problem 4.3

For each of the following systems of simultaneous linear congruences decide whether or not it has solutions and if so solve the system.

(a) $x \equiv 7 \;(\text{mod } 12)\,; \quad x \equiv 11 \;(\text{mod } 18)\,; \quad x \equiv 1 \;(\text{mod } 23)$

(b) $x \equiv 4 \;(\text{mod } 12)\,; \quad x \equiv 10 \;(\text{mod } 18)\,; \quad x \equiv 3 \;(\text{mod } 7)$

Throughout the remainder of the course we shall continually be meeting congruence problems. It may be surprising how often the task of solving linear congruences crops up in the context of deeper problems.

ADDITIONAL EXERCISES

Section 1

1 Determine the least positive residue and the least absolute residue of 1997

   (a) modulo 7;

   (b) modulo 17;

   (c) modulo 101.

2 Given that $k \equiv 1 \;(\text{mod } 12)$, what is the least positive residue of $6k + 5$

   (a) modulo 4;

   (b) modulo 9;

   (c) modulo 12?

3 (a) Give an example to show that $a^k \equiv b^k \;(\text{mod } n)$ need not imply that $a \equiv b \;(\text{mod } n)$.

   (b) Give an example to show that $a^k \equiv b^k \;(\text{mod } n)$ and $k \equiv j \;(\text{mod } n)$ both being true does not necessarily imply that $a^j \equiv b^j \;(\text{mod } n)$ is true.

4 Prove by induction that $6^n \equiv 5n + 1 \;(\text{mod } 25)$, for all integers $n \geq 1$.

5 Prove that if $n \geq 2$ and $\gcd(a, n) = 1$ then the set of integers

   $$\{c, c + a, c + 2a, c + 3a, \ldots, c + (n-1)a\}$$

   is a complete set of residues modulo n for any integer c.

6   What is the remainder when 11^{11} is divided by

    (a) 7;

    (b) 17;

    (c) 19.

7   Use ideas of congruence to prove that

    (a) 29 divides $2^{28} - 1$;

    (b) 79 divides $3^{39} + 1$.

8   In 1659 Fermat asserted that no prime number of the form $3k - 1$ can be written in the form $a^2 + 3b^2$. Use congruence modulo 3 to show that he was correct.

9   From the definition of congruence alone, prove the following.

    If $ar \equiv b \pmod{n}$ and $cr \equiv d \pmod{n}$, for some integer r, then $ad \equiv bc \pmod{n}$.

Section 2

1   Without carrying out the multiplication find the missing digit (?) in each of the following products.

    (a) $108 \times 371 = 40\,0(?)8$

    (b) $1578 \times 34\,971 = 55\,18(?)\,238$

    (c) $4473 \times 3(?)9 = 1\,426\,887$ *Hint*: 11 divides $1\,426\,887$.

2 A and B are two different 7-digit numbers each involving all seven of the digits 1, 2, 3, 4, 5, 6 and 7. Prove that A does not divide B.

3   Let $N > M$ be two 10-digit numbers formed from the same 10 digits but in different orders. If X is the sum of the digits of $N - M$, show that the sum of the digits of X is 9.

4   Show that 2^n divides an integer N if, and only if, 2^n divides the number made up of the last n digits of N.

5   (a) Show that, for all $n \geq 0$, $10^{2n} \equiv 1 \pmod{99}$ and $10^{2n+1} \equiv 10 \pmod{99}$.

    (b) The remainder on dividing a number by 99 can be found in a way similar to the corresponding result for division by 9, but treating the digits of the number two at a time. The following examples illustrate the method:

$$6239 \equiv 62 + 39 \equiv 101 \equiv 2 \pmod{99}$$

    and

$$54\,787 \equiv 5 + 47 + 87 \equiv 139 \equiv 40 \pmod{99}.$$

    Use the result in part (a) to show why this test works.

    (c) Find a test for divisibility by 101 and use it to determine the remainders when $62\,479$ and $9\,850\,833$ are divided by 101.

Section 3

1 Solve the following linear congruences.

    (a) $8x \equiv 12 \pmod{60}$

    (b) $21x \equiv 35 \pmod{47}$

    (c) $6x \equiv 5 \pmod{41}$

    (d) $5x \equiv 1 \pmod{93}$

2 Solve the following polynomial congruences.

    (a) $x^3 \equiv 1 \pmod{7}$

    (b) $2x^2 + 4x + 5 \equiv 0 \pmod{7}$

3 Use congruences to solve the following linear Diophantine equations.

    (a) $12x + 19y = 7$

    (b) $17x + 41y = 13$

4 There are 19×20 congruences of the form $ax \equiv b \pmod{20}$, since there are 19 different choices for $a \not\equiv 0 \pmod{20}$ and 20 choices for b. For how many of these congruences does a solution exist, and for how many congruences is the solution unique?

Section 4

1 Solve the following system of congruences by following the steps in the proof of the Chinese Remainder Theorem.

$$x \equiv 1 \pmod{4}; \quad x \equiv 3 \pmod{5}; \quad x \equiv 0 \pmod{7}.$$

2 Solve the following systems of congruences by any method.

    (a) $2x \equiv 3 \pmod{5}; \quad 3x \equiv 4 \pmod{7}; \quad 4x \equiv 9 \pmod{11}$.

    (b) $x \equiv 1 \pmod{4}; \quad x \equiv 3 \pmod{6}; \quad x \equiv 3 \pmod{9}$.

3 Solve the following linear congruence.

$$19x \equiv 7 \pmod{2310}$$

4 Find the smallest positive integer n such that 3^2 divides n, 4^2 divides $n + 1$ and 5^2 divides $n + 2$.

5 In the arithmetic progression

$$7, \ 18, \ 29, \ 40, \ 51, \ \ldots$$

find the first three consecutive terms which are divisible by 2, 3 and 5 respectively.

6 For which values of c does the following system of congruences have a simultaneous solution?

$$5x \equiv c \pmod{12}; \quad 7x \equiv 2 \pmod{30}$$

7 A band of seventeen pirates stole a sack of gold coins. Attempting to divide the spoils into equal portions they found that four coins were left over. A brawl over who should get the extra coins followed and one pirate was killed. When the spoils were redistributed amongst the survivors, ten coins were left over. Again an argument developed and one pirate was killed. But now the spoils divided equally between the remaining pirates. What was the least number of coins that could have been stolen?

Challenge Problems

1 Let $A = 4444^{4444}$. If B is the sum of the digits of A, C is the sum of the digits of B and D is the sum of the digits of C, what is D?

2 Find a positive integer such that one-half of it is a square, one-third of it is a cube and one-fifth of it is a fifth power.

3 Find the final digit of each of the numbers

$$(\ldots(((7^7)^7)^7)^7 \ldots)^7 \quad \text{and} \quad 7^{\left(7^{\cdots^{\left(7^7\right)}}\right)}$$

where, in both expressions, the integer 7 occurs $1\,001$ times. What are the final two digits of each?

4 Let N be any integer. How can you construct from the digits of N a number M of at most 3 digits such that $N \equiv M \pmod{37}$.

SOLUTIONS TO THE PROBLEMS

Solution 1.1

(a) False, as $3 - 29 = -26$, which is not a multiple of 8.

(b) True, as $3 - (-29) = 32 = 4 \times 8$.

(c) True, as $63 - 37 = 26 = 2 \times 13$.

(d) True, as $37 - 63 = -26 = (-2) \times 13$.

(e) True, as $63 - 63 = 0 = 0 \times 37$.

(f) False, as $63 - 13 = 50 = 2 \times 25$, so $63 \equiv 13 \pmod{25}$.

> From the definition of divisors (Definition 4.1 of *Unit 1*) 0 is divisible by every positive integer.

Solution 1.2

Since $a \equiv b \pmod{n}$, let k be the integer such that $a - b = kn$.

(a) m divides n means that $n = mr$ for some integer r. Therefore

$$a - b = kn = k(mr) = (kr)m,$$

which confirms that $a \equiv b \pmod{m}$.

(b) Multiplying by m:

$$ma - mb = mkn = k(mn), \quad \text{confirming } ma \equiv mb \pmod{mn},$$
$$= (mk)n, \quad \text{confirming } ma \equiv mb \pmod{n}.$$

(c) Dividing by d:

$$\frac{a}{d} - \frac{b}{d} = k\frac{n}{d}, \quad \text{which implies that} \quad \frac{a}{d} \equiv \frac{b}{d} \left(\text{mod } \frac{n}{d}\right).$$

> Note that $\frac{a}{d}, \frac{b}{d}$ and $\frac{n}{d}$ are integers.

(d) Let $\gcd(a, n) = d_1$ and $\gcd(b, n) = d_2$. Since d_1 divides a, and d_1 divides n, it follows that d_1 divides $a - kn$; that is d_1 divides b. But then d_1 is a common divisor of b and n, and so $d_1 \leq d_2$.

On the other hand, d_2 divides b, and d_2 divides n, giving d_2 divides $b + kn$, that is, d_2 divides a. So d_2 is a common divisor of a and n, and so $d_2 \leq d_1$. Hence the equality $d_1 = d_2$ follows.

(e) $(a + c) - (b + c) = a - b = kn,$ and so $a + c \equiv b + c \pmod{n}.$

Note that the result in part (e) is just a special, but important, case of Theorem 1.1(d) in which $d = c$.

Solution 1.3

(a) The set of least positive residues modulo 10 is

$$\{0, 1, 2, 3, 4, 5, 6, 7, 8, 9\}.$$

The set of least absolute residues modulo 10 is

$$\{-4, -3, -2, -1, 0, 1, 2, 3, 4, 5\}.$$

(b) The set of least positive residues modulo 11 is

$$\{0, 1, 2, 3, 4, 5, 6, 7, 8, 9, 10\}.$$

The set of least absolute residues modulo 11 is

$$\{-5, -4, -3, -2, -1, 0, 1, 2, 3, 4, 5\}.$$

Solution 1.4

(a) Replacing each of the numbers by the least positive residue to which it is congruent modulo 11, (for instance $-3 \equiv 8, 22 \equiv 0$, etc), we obtain

$$\{6, 2, -3, 22, 23, -7, 7, 9, 43, -8, 16\} \equiv \{6, 2, 8, 0, 1, 4, 7, 9, 10, 3, 5\}.$$

As the latter is the set of least positive residues modulo 11 the former is a complete set of residues modulo 11.

When there is no ambiguity over which modulus is involved we shall occasionally, as in this solution, use the $\equiv$ symbol without its accompanying $\pmod{n}$.

(b) As $10^2 = 100 \equiv 1$ and $1^2 \equiv 1 \pmod{11}$ the eleven numbers do not form a complete set of residues. In fact only six residue classes are represented in this set, as $9^2 \equiv 2^2$, $8^2 \equiv 3^2$, $7^2 \equiv 4^2$ and $6^2 \equiv 5^2$.

(c) Starting from 1, each subsequent number is obtained by doubling. To avoid calculation of high powers of 2 we can write down the least positive residue for each of the numbers by systematically 'doubling modulo 11'. The given eleven numbers are congruent modulo 11, in order, to the following:

$$0, 1, 2, 4, 8, 16 \equiv 5, 10, 20 \equiv 9, 18 \equiv 7, 14 \equiv 3 \text{ and } 6.$$

As these numbers form the set of least positive residues modulo 11, the given set is a complete set of residues.

Each time 11 is exceeded the number is replaced by the least positive residue and doubling begins again.

Solution 1.5

In $\mathbb{Z}_{10}$:

$$[6] + [8] = [6 + 8] = [14] = [4], \text{ since } 14 \equiv 4 \pmod{10}.$$
$$[6] \times [8] = [6 \times 8] = [48] = [8], \text{ since } 48 \equiv 8 \pmod{10}.$$

Allowing b to take each of the values $0, 1, 2, \ldots, 8$ and 9 in turn we find that the corresponding values of $[6] \times [b]$ are $[0], [6], [2], [8], [4], [0], [6], [2], [8],$ and $[4]$. So $[6] \times [b] = [8]$ has the two solutions $[b] = [3]$ and $[b] = [8]$.

Repeating in $\mathbb{Z}_{11}$:

$$[6] + [8] = [14] = [3], \text{ since } 14 \equiv 3 \pmod{11}.$$
$$[6] \times [8] = [48] = [4], \text{ since } 48 \equiv 4 \pmod{11}.$$
$$[6] \times [b] = [8] \text{ has the unique solution } [b] = [5].$$

Solution 1.6

We first note that if $n \geq 7$ then 7 divides $n!$ and so $n! \equiv 0$ (mod 7). Therefore we have the simplification:

$$1! + 2! + 3! + \cdots + 1000! \equiv 1! + 2! + 3! + \cdots + 6! \text{ (mod 7)}$$

(which removes all fear of the problem!) Now, working modulo 7, $1! = 1$, $2! = 2$, $3! = 6$, $4! = 24 \equiv 3$, $5! = 5 \times 4! \equiv 5 \times 3 \equiv 1$ and $6! = 6 \times 5! \equiv 6 \times 1 = 6$. And so,

$$1! + 2! + 3! + \cdots + 1000! \equiv 1 + 2 + 6 + 3 + 1 + 6 = 19 \equiv 5 \text{ (mod 7)},$$

and the remainder on dividing by 7 is 5.

Solution 1.7

$a^2 - 1$ is divisible by 24 precisely when $a^2 - 1 \equiv 0$ (mod 24) or equivalently, adding 1 to both sides and appealing to Theorem 1.1(d), when $a^2 \equiv 1$ (mod 24). So our task is to show that any integer, a, which is divisible neither by 2 nor by 3 must satisfy $a^2 \equiv 1$ (mod 24).

The Division Algorithm tells us that a can be written as

$$a = 24q + r, \text{ where } 0 \leq r \leq 23.$$

But, written in this way, a is divisible by 2 when r is even, and is divisible by 3 when r is divisible by 3. There are just eight values of r which are not divisible by either 2 or 3, namely $r = 1, 5, 7, 11, 13, 17, 19$ and 23. For a to be not divisible by either 2 or 3 it must be congruent modulo 24 to one of these eight values. Thus we have, replacing each of the last four listed values by its least absolute residue,

$$a \equiv \pm 1, \pm 5, \pm 7, \pm 11 \text{ (mod 24)}.$$

Now appealing to Theorem 1.1(f),

$$a^2 \equiv 1, 25, 49 \text{ or } 121 \text{ (mod 24)}.$$

But each of these numbers has remainder 1 when divided by 24, and the result follows.

Solution 1.8

(a) We wish to cancel the common divisor 6. As $\gcd(6, 30) = 6$, we also divide the modulus by 6:

$$6x \equiv 18 \text{ (mod 30)} \iff x \equiv 3 \text{ (mod 5)}.$$

(b) In order to cancel the common divisor 10, the modulus has to be divided by $\gcd(10, 45) = 5$. So,

$$20x \equiv 30 \text{ (mod 45)} \iff 2x \equiv 3 \text{ (mod 9)}.$$

(c) The common divisor 13 can be cancelled without adjustment to the modulus, since $\gcd(13, 60) = 1$. So

$$52x \equiv 39 \text{ (mod 60)} \iff 4x \equiv 3 \text{ (mod 60)}.$$

Solution 2.1

(a) (i)
$$N \equiv 9 + 7 + 2 + 3 \equiv 3 \text{ (mod 9)};$$
$$M \equiv 1 + 8 + 0 + 5 + 6 \equiv 2 \text{ (mod 9)};$$
$$N + M \equiv 3 + 2 \equiv 5 \text{ (mod 9)};$$
$$N - M \equiv 3 - 2 \equiv 1 \text{ (mod 9)}.$$

Hence the required remainders are 3, 2, 5 and 1 respectively.

(ii)
$$N \equiv 3 - 2 + 7 - 9 \equiv -1 \equiv 10 \ (\text{mod } 11) \, ;$$
$$M \equiv 6 - 5 + 0 - 8 + 1 \equiv -6 \equiv 5 \ (\text{mod } 11) \, ;$$
$$N + M \equiv 10 + 5 \equiv 4 \ (\text{mod } 11) \, ;$$
$$N - M \equiv 10 - 5 \equiv 5 \ (\text{mod } 11) \, .$$

Hence the required remainders are 10, 5, 4 and 5 respectively.

(b) $N \times M \equiv 3 \times 2 \equiv 6 \ (\text{mod } 9)$, and so the number on the right-hand side of the equation has to be congruent modulo 9 to 6.

$$1 + 7 + 5 + (?) + 5 + 8 + 4 + 8 + 8 = 46 + (?) \equiv 1 + (?) \equiv 6 \ (\text{mod } 9) \, ,$$

from which it follows that the missing digit (?) can only be 5.

Solution 2.2

Let $N = a_{2m-1} a_{2m-2} \ldots a_2 a_1 a_0$ so that $M = a_0 a_{2m-1} a_{2m-2} \ldots a_2 a_1$. Since N and M comprise the same $2m$ digits, the 9-test gives

$$N \equiv M \equiv a_0 + a_1 + a_2 + \cdots + a_{2m-2} + a_{2m-1} \ (\text{mod } 9) \, ,$$

and so $N - M \equiv 0 \ (\text{mod } 9)$, that is, 9 divides $N - M$.

Applying the 11-test:

$$N \equiv a_0 - a_1 + a_2 - \cdots + a_{2m-2} - a_{2m-1} \ (\text{mod } 11)$$

whilst

$$M \equiv a_1 - a_2 + a_3 \cdots - a_{2m-2} + a_{2m-1} - a_0 \ (\text{mod } 11) \, .$$

So $N + M \equiv 0 \ (\text{mod } 11)$. That is, 11 divides $N + M$.

Finally, since $N^2 - M^2 = (N - M)(N + M)$ we have that 9 divides $N^2 - M^2$ and 11 divides $N^2 - M^2$. As $\gcd(9, 11) = 1$, the Corollary to Theorem 1.3 allows us to conclude that 99 divides $N^2 - M^2$.

Solution 3.1

(a) We evaluate $P_1(x)$ for each of the least positive residues modulo 7, namely $x = 0, 1, 2, 3, 4, 5$ and 6.

$$0^2 + 3 \times 0 + 4 = 4 \equiv 4 \ (\text{mod } 7)$$
$$1^2 + 3 \times 1 + 4 = 8 \equiv 1 \ (\text{mod } 7)$$
$$2^2 + 3 \times 2 + 4 = 14 \equiv 0 \ (\text{mod } 7)$$
$$3^2 + 3 \times 3 + 4 = 22 \equiv 1 \ (\text{mod } 7)$$
$$4^2 + 3 \times 4 + 4 = 32 \equiv 4 \ (\text{mod } 7)$$
$$5^2 + 3 \times 5 + 4 = 44 \equiv 2 \ (\text{mod } 7)$$
$$6^2 + 3 \times 6 + 4 = 58 \equiv 2 \ (\text{mod } 7)$$

So $P_1(x) \equiv 0 \ (\text{mod } 7)$ has the one solution $x \equiv 2 \ (\text{mod } 7)$.

(b) We need to evaluate $P_2(x)$ on a complete set of residues. This time we shall choose the least absolute residues modulo 5, namely $x = -2, -1, 0, 1$ and 2.

$$P_2(0) = -1 \equiv 4 \ (\text{mod } 5)$$
$$P_2(1) = P_2(-1) = 1 - 1 \equiv 0 \ (\text{mod } 5)$$
$$P_2(2) = P_2(-2) = 16 - 1 \equiv 0 \ (\text{mod } 5)$$

The fact that $P(a) = P(-a)$ influences the use of the least absolute residues for, as you see, it effectively reduces the number of residues to be checked.

$P_2(x) \equiv 0 \ (\text{mod } 5)$ has the four solutions $x \equiv 1, 2, -2$ and $-1 \ (\text{mod } 5)$. That is, alternatively, $x \equiv 1, 2, 3$ and 4 (modulo 5).

Solution 3.2

(a) Since $\gcd(5, 10) = 5$ and 5 does not divide 13, this congruence has *no* solutions.

(b) Since $\gcd(5, 13) = 1$, this congruence has a *unique* solution.

(c) Since $\gcd(10, 30) = 10$ and 10 does not divide 5, this congruence has *no* solutions.

(d) Since $\gcd(27, 24) = 3$ and 3 divides 318, this congruence has three solutions.

Solution 3.3

(a) The congruence has a unique solution modulo 53.

$$28x \equiv 49 \pmod{53} \iff 4x \equiv 7 \pmod{53} \quad \text{Step 3}$$
$$\iff 4x \equiv 60 \pmod{53} \quad \text{Step 4}$$
$$\iff x \equiv 15 \pmod{53} \quad \text{Step 3}$$

(b) $\gcd(34, 52) = 2$ and 2 does not divide 19 so the congruence has *no* solutions.

(c) The congruence has a unique solution modulo 19.

$$137x \equiv 96 \pmod{19} \iff 4x \equiv 1 \pmod{19} \quad \text{Step 4}$$
$$\iff 4x \equiv 20 \pmod{19} \quad \text{Step 4}$$
$$\iff x \equiv 5 \pmod{19} \quad \text{Step 3}$$

$137 \equiv 4 \pmod{19}$ and $96 \equiv 1 \pmod{19}$.

(d) The congruence has a unique solution modulo 53.

$$51x \equiv 8 \pmod{53} \iff -2x \equiv 8 \pmod{53} \quad \text{Step 4}$$
$$\iff 2x \equiv -8 \pmod{53} \quad \text{Step 5}$$
$$\iff x \equiv -4 \pmod{53} \quad \text{Step 3}$$
$$\iff x \equiv 49 \pmod{53} \quad \text{Step 4}$$

Rather than multiply through by -1 and then divide by 2 we could simply divide through by -2.

(e) As $\gcd(18, 69) = 3$ and 3 divides 39, this congruence has three solutions modulo 69.

$$18x \equiv 39 \pmod{69} \iff 6x \equiv 13 \pmod{23} \quad \text{Step 2}$$
$$\iff 6x \equiv 36 \pmod{23} \quad \text{Step 4}$$
$$\iff x \equiv 6 \pmod{23} \quad \text{Step 3}$$
$$\iff x \equiv 6, 29, 52 \pmod{69}$$

We still use the equivalence symbol $\iff$ despite the different moduli, because an integer x is a solution of the previous congruence if, and only if, it is a solution of the following congruence.

The final step gives the derived solution in terms of the original modulus.

Solution 4.1

Expressed in terms of congruence the three conditions to be satisfied are

$$2n + 3 \equiv 0 \pmod{9}; \quad 3n + 4 \equiv 0 \pmod{16}; \quad 4n + 5 \equiv 0 \pmod{25}.$$

Solving these individually we have the equivalent system

$$n \equiv 3 \pmod{9}; \quad n \equiv 4 \pmod{16}; \quad n \equiv 5 \pmod{25}.$$

The variables in the proof of the Chinese Remainder Theorem are:
$b_1 = 3$, $b_2 = 4$, $b_3 = 5$, $n_1 = 9$, $n_2 = 16$, $n_3 = 25$, $N = 3600$, $N_1 = 400$, $N_2 = 225$, $N_3 = 144$.

$$N_1 x = 400x \equiv 1 \pmod{9} \iff 4x \equiv 1 \pmod{9}$$
$$\iff x \equiv 7 \pmod{9}, \text{ so } x_1 = 7.$$
$$N_2 x = 225x \equiv 1 \pmod{16} \iff x \equiv 1 \pmod{16}, \text{ so } x_2 = 1.$$
$$N_3 x = 144x \equiv 1 \pmod{25} \iff 19x \equiv 1 \equiv 76 \pmod{25}$$
$$\iff x \equiv 4 \pmod{25}, \text{ so } x_3 = 4.$$

Therefore
$$x_0 = 3 \times 400 \times 7 + 4 \times 225 \times 1 + 5 \times 144 \times 4 = 12\,180 \equiv 1380 \pmod{3600}$$

and the least positive solution is $n = 1380$.

Solution 4.2

As $165 = 3 \times 5 \times 11$ we begin by solving the linear congruence to each of the moduli 3, 5 and 11.

$$7x \equiv 9 \ (\text{mod } 3) \iff x \equiv 0 \ (\text{mod } 3)$$
$$7x \equiv 9 \ (\text{mod } 5) \iff x \equiv 2 \ (\text{mod } 5)$$
$$7x \equiv 9 \ (\text{mod } 11) \iff x \equiv 6 \ (\text{mod } 11)$$

Our task is now to solve the three congruences on the right-hand side simultaneously.

$$x \equiv 6 \ (\text{mod } 11) \implies x = 6, 17, \ldots,$$

increasing in steps of 11 until we reach a solution of the second congruence.

$$x \equiv 2 \ (\text{mod } 5) \implies x = 17, 72, \ldots,$$

increasing in steps of $11 \times 5 = 55$ until we reach a solution of the remaining congruence.

$$x \equiv 0 \ (\text{mod } 3) \implies x \equiv 72 \ (\text{mod } 165).$$

Solution 4.3

(a) This system has no solutions since $\gcd(12, 18) = 6$ and 6 does not divide $11 - 7 = 4$.

(b) This system has solutions since $\gcd(12, 18)$ divides $10 - 4$, $\gcd(7, 12) = 1$ (which divides all integers) and $\gcd(7, 18) = 1$.

$$x \equiv 10 \ (\text{mod } 18) \implies x = 10, 28, \ldots,$$
$$x \equiv 4 \ (\text{mod } 12) \implies x = 28, 64, 100, 136, \ldots,$$
$$\text{increasing in steps of } \text{lcm}(12, 18) = 36,$$
$$x \equiv 3 \ (\text{mod } 7) \implies x \equiv 136, 388, 640, \ldots,$$
$$\text{increasing in steps of } 7 \times 36 = 252.$$

The solution $x \equiv 136 \ (\text{mod } 252)$ is unique.

SOLUTIONS TO ADDITIONAL EXERCISES

Section 1

1   (a) Since $1997 = 7 \times 285 + 2$,
the least positive residue is 2, and
the least absolute residue is 2.

(b) Since $1997 = 17 \times 117 + 8$,
the least positive residue is 8, and
the least absolute residue is 8.

(c) Since $1997 = 101 \times 19 + 78$,
the least positive residue is 78, and
the least absolute residue is $78 - 101 = -23$.

2 If $k \equiv 1 \pmod{12}$ then $k = 12r + 1$, for some integer r. Hence
$6k + 5 = 72r + 11$.

(a) As 4 divides 72, we have $6k + 5 \equiv 11 \equiv 3 \pmod 4$, and so 3 is the required least positive residue.

(b) As 9 divides 72, we have $6k + 5 \equiv 11 \equiv 2 \pmod 9$, and so 2 is the required least positive residue.

(c) As 12 divides 72, we have $6k + 5 \equiv 11 \pmod{12}$, and so 11 is the required least positive residue.

3 (a) $1^2 \equiv 6^2 \pmod 7$ but it is not the case that $1 \equiv 6 \pmod 7$.

(b) $1^2 \equiv 2^2 \pmod 3$ and $2 \equiv 5 \pmod 3$, but $1^5 \not\equiv 2^5 \pmod 3$.

4 Let $P(n)$ be the proposition $6^n \equiv 5n + 1 \pmod{25}$.

$P(1)$ claims that $6 \equiv 6 \pmod{25}$ which is trivially true. We continue by assuming that $P(k)$ is true and investigate $P(k+1)$.

$$6^{k+1} = 6 \times 6^k \equiv 6 \times (5k + 1), \quad \text{by the induction hypothesis,}$$
$$\equiv 30k + 6 \pmod{25}$$
$$\equiv 5k + 6 \pmod{25}$$
$$\equiv 5(k + 1) + 1 \pmod{25},$$

as required.

The Principle of Mathematical Induction confirms that $P(n)$ is true for all integers $n \geq 1$.

5 The set of n numbers will form a complete set of residues modulo n provided no two of them are congruent modulo n. So, aiming for a contradiction, suppose that two of the numbers are congruent modulo n, say

$$c + ra \equiv c + sa \pmod n, \quad \text{where } 0 \leq s < r < n.$$

The definition of congruence would then give

$$n \text{ divides } (c + ra) - (c + sa).$$

That is,

$$n \text{ divides } (r - s)a.$$

But $\gcd(a, n) = 1$, and so Euclid's Lemma leads to

$$n \text{ divides } r - s,$$

which is impossible since $0 < r - s < n$.

6 There is no unique set of steps to answer these questions. You might well have found different ways of grouping the terms from those in the solutions which follow.

(a) As $11 \equiv 4 \pmod 7$, we have $11^{11} \equiv 4^{11} \pmod 7$.

Now $4^2 = 16 \equiv 2 \pmod 7$ and so

$$11^{11} \equiv 4^{11} \equiv (4^2)^5 \times 4 \equiv 2^5 \times 4 \equiv 32 \times 4 \equiv 4 \times 4 \equiv 2 \pmod 7.$$

So the remainder is 2.

(b) As $11^2 = 121 \equiv 2 \pmod{17}$, we have

$$11^{11} = (11^2)^5 \times 11 \equiv 2^5 \times 11 \equiv (-2) \times 11 \equiv -5 \equiv 12 \pmod{17}.$$

So the remainder is 12.

(c) $11^{11} \equiv (-8)^{11} \equiv ((-8)^2)^5 \times (-8) \equiv 7^5 \times (-8) \pmod{19}$

$\equiv 7^2 \times 7^2 \times 7 \times (-8) \equiv 11 \times 11 \times (-56) \pmod{19}$

$\equiv (-8)^2 \times 1 \equiv 64 \equiv 7 \pmod{19}.$

So the remainder is 7.

7 (a) $2^{28} \equiv (2^5)^5 \times 2^3 \equiv 3^5 \times 2^3 \equiv 27 \times 9 \times 8 \pmod{29}$

$\equiv (-2) \times 14 \equiv -28 \equiv 1 \pmod{29},$

which amounts to saying that 29 divides $2^{28} - 1$.

(b) $3^{39} \equiv (3^4)^9 \times 3^3 \equiv 2^9 \times 3^3 \equiv (8 \times 9) \times (32 \times 3) \times 2 \equiv (-7) \times (17) \times 2 \pmod{79}$

$\equiv (-119) \times 2 \equiv (-40) \times 2 \equiv -80 \equiv -1 \pmod{79}.$

It follows that 79 divides $3^{39} + 1$.

8 Since any square is congruent modulo 3 to either 0 or 1 it follows that $a^2 + 3b^2$ is congruent modulo 3 to either 0 or 1 (since $3b^2 \equiv 0 \pmod 3$). But $3k - 1 \equiv 2 \pmod 3$, and so $a^2 + 3b^2$ cannot be equal to $3k - 1$ for any integer k, regardless of whether or not it is prime.

9 The congruences $ar \equiv b \pmod n$ and $cr \equiv d \pmod n$ tell us that there exist integers k and s such that $ar - b = kn$ and $cr - d = sn$. Substituting for b from the first of these equations and for d from the second gives

$$ad - bc = a(cr - sn) - (ar - kn)c = n(kc - as),$$

which shows that $ad \equiv bc \pmod n$.

Section 2

1 (a) As 108 is divisible by 9, the right-hand side must be divisible by 9. Hence, by adding up the digits, $4 + 0 + 0 + (?) + 8 \equiv 0 \pmod 9$ and the only value of the digit $(?)$ which renders this possible is 6.

(b) Checking the calculation modulo 9,

$$1578 \times 34\,971 \equiv 3 \times 6 \equiv 0 \pmod 9,$$

and so the sum of the digits on the right-hand side must be divisible by 9. Thus

$$5 + 5 + 1 + 8 + (?) + 2 + 3 + 8 \equiv 32 + (?) \equiv 5 + (?) \equiv 0 \pmod 9,$$

giving $(?) = 4$.

(c) The right-hand side is divisible by 11 and so the left-hand side must be divisible by 11. Now 4473 is not divisible by 11, and so $3(?)9$ must be. Hence $(?) = 1$ is the only possibility.

2 Aiming for a contradiction, suppose that A divides B, say $B = nA$ for some integer n. Now as A comprises the digits 1 to 7 we have

$$A \equiv 1 + 2 + 3 + 4 + 5 + 6 + 7 = 28 \equiv 1 \pmod 9$$

and similarly $B \equiv 1 \pmod 9$.

The equation $B = nA$ then gives $n \equiv 1 \pmod 9$. But $n = 1$ gives $B = A$, which is not the case, and so $n \geq 10$. However this is impossible since that forces B to have more digits than A. This contradiction shows that no such n can exist and so A does not divide B.

3 The key here is to note that, as M and N have the same collection of digits, $N \equiv M \pmod 9$ or, what amounts to the same thing, $N - M \equiv 0 \pmod 9$. But X is the sum of the digits of $N - M$ and so $X \equiv 0 \pmod 9$. Now $N - M$ is a number with ten (or fewer) digits and so $X \leq 90$. Of all numbers not exceeding 90, the maximum sum of the digits is 17 (achieved by the number 89). X cannot be equal to 0 as that would happen if, and only if, $N = M$.

Let Y be the sum of the digits of X. Then

$$Y \equiv X \equiv 0 \pmod 9, \quad \text{where } 0 < Y < 18.$$

Thus the only possibility is $Y = 9$.

4 Dividing N, by 10^n, the Division Algorithm gives integers q and r such that

$$N = 10^n q + r, \quad 0 \leq r < 10^n.$$

Because of the restriction on r, it follows that r is the number made up of the last n digits of N.

Now 2^n divides 10^n and so, taking congruence modulo 2^n,

$$N \equiv r \pmod{2^n}.$$

Thus $N \equiv 0 \pmod{2^n}$ if, and only if, $r \equiv 0 \pmod{2^n}$, that is, 2^n divides N if, and only if, 2^n divides r, the number made up of the last n digits of N.

For example, to extract the final three digits from $1\,234\,567$ we write $1\,234\,567 = 1234 \times 10^3 + 567$.

5 (a) $10^{2n} = (10^2)^n \equiv 1^n \equiv 1 \pmod{99}$,

$\qquad 10^{2n+1} = 10 \times 10^{2n} \equiv 10 \pmod{99}$.

(b) Let

$$N = a_m a_{m-1} \ldots a_2 a_1 a_0$$
$$= a_0 + a_1 10 + a_2 10^2 + a_3 10^3 + \cdots + a_{m-1} 10^{m-1} + a_m 10^m$$
$$= (10a_1 + a_0) + 10^2 (10a_3 + a_2) + 10^4 (10a_5 + a_4) + \ldots.$$

The last term in the summation is $10^{m-1}(10a_m + a_{m-1})$ if m is odd and $10^m(0 + a_m)$ if m is even.

Utilizing the fact that $10^{2n} = 1 \pmod{99}$ we have

$$N \equiv (10a_1 + a_0) + (10a_3 + a_2) + (10a_5 + a_4) + \ldots \pmod{99},$$

and the remainder on dividing by 99 can be obtained from this sum of pairs of digits.

(c) Working modulo 101, the successive powers of 10 are:

$$10^0 \equiv 1, 10^1 \equiv 10, 10^2 \equiv -1, 10^3 \equiv -10, 10^4 \equiv 1, 10^5 \equiv 10, \text{ etc.}$$

Hence

$$N = a_m a_{m-1} \ldots a_2 a_1 a_0$$
$$= a_0 + a_1 10 + a_2 10^2 + a_3 10^3 + \cdots + a_{m-1} 10^{m-1} + a_m 10^m$$
$$\equiv (10a_1 + a_0) - (10a_3 + a_2) + (10a_5 + a_4) - \cdots \pmod{101},$$

and the remainder on dividing by 101 is obtained from this alternating sum of pairs of digits. Now

$$62\,479 \equiv 79 - 24 + 6 \equiv 61 \pmod{101} \text{ and}$$
$$9\,850\,833 \equiv 33 - 8 + 85 - 9 \equiv 101 \equiv 0 \pmod{101},$$

so the remainders are 61 and 0 respectively.

Section 3

1   (a) $8x \equiv 12 \pmod{60} \iff 2x \equiv 3 \pmod{15}$
$$\iff 2x \equiv 18 \pmod{15}$$
$$\iff x \equiv 9 \pmod{15}$$
$$\iff x \equiv 9, 24, 39, 54 \pmod{60}$$

   (b) $21x \equiv 35 \pmod{47} \iff 3x \equiv 5 \pmod{47}$
$$\iff 3x \equiv -42 \pmod{47}$$
$$\iff x \equiv -14 \equiv 33 \pmod{47}$$

   (c) $6x \equiv 5 \pmod{41} \iff 42x \equiv 35 \pmod{41}$
$$\iff x \equiv 35 \pmod{41}$$

   (d) $5x \equiv 1 \pmod{93} \iff 95x \equiv 19 \pmod{93}$
$$\iff 2x \equiv 112 \pmod{93}$$
$$\iff x \equiv 56 \pmod{93}$$

2   (a) Working modulo 7 gives:

$$0^3 \equiv 0, \; 1^3 \equiv 1, \; 2^3 \equiv 1, \; 3^3 \equiv 6, \; 4^3 \equiv 1, \; 5^3 \equiv 6 \text{ and } 6^3 \equiv 6.$$

Therefore the equation $x^3 \equiv 1 \pmod 7$ has solutions $x \equiv 1, 2, 4 \pmod 7$.

   (b) The value of $2x^2 + 4x + 5 \pmod 7$ for each of the seven values of x in a complete set of residues is given in the following table.

x	0	1	2	3	4	5	6
$2x^2 + 4x + 5$	5	4	0	0	4	5	3

Hence $2x^2 + 4x + 5 \equiv 0$ has the two solutions $x \equiv 2, 3 \pmod 7$.

3   (a) $12x + 19y = 7$.

The obvious moduli to try are 12 and 19 as each of these eliminates one of the variables.

Writing this equation as a congruence modulo 12:

$$19y \equiv 7 \pmod{12},$$

which gives

$$y \equiv 1 \pmod{12}.$$

Substituting $y = 1$ in the original equation gives $x = -1$, and so the general solution is

$$x = -1 + 19k, \quad y = 1 - 12k, \quad k \in \mathbb{Z}.$$

   (b) $17x + 41y = 13$.

Writing this equation modulo 17 we have

$$41y \equiv 13 \pmod{17} \iff 7y \equiv 13 \pmod{17}$$
$$\iff 35y \equiv 65 \pmod{17} \quad \text{multiplying by 5}$$
$$\iff y \equiv 14 \equiv -3 \pmod{17}.$$

Substituting $y = -3$ in the original equation gives $x = 8$, and hence the general solution is

$$x = 8 + 41k, \quad y = -3 - 17k, \quad k \in \mathbb{Z}.$$

4   Take the uniqueness first. The congruence $ax \equiv b \pmod{20}$ has a unique solution if, and only if, $\gcd(a, 20) = 1$; that is, when $a = 1, 3, 7, 9, 11, 13, 17$ and 19. For each of these eight values of a there are twenty values of b, making 160 congruences in total for which the solution is unique.

The congruence has $\gcd(a, 20)$ solutions whenever $\gcd(a, 20)$ divides b. So consider the possible values of $\gcd(a, 20)$ in turn.

$\gcd(a, 20) = 1$ leads to 160 solvable congruences, as already seen.

$\gcd(a, 20) = 2$ when $a = 2, 6, 14$ and 18. For each of these four values of a the congruence is solvable provided b takes one of the ten even values $0, 2, 4, 6, 8, 10, 12, 14, 16$ or 18. Hence there are 40 congruences with solutions.

$\gcd(a, 20) = 4$ when $a = 4, 8, 12$ and 16. For each of these four values of a the congruence has solutions when b takes one of the five values $0, 4, 8, 12$ or 16 making 20 congruences in all.

$\gcd(a, 20) = 5$ when $a = 5$ or 15. For each of these two values of a the congruence has solutions if b takes one of the values $0, 5, 10$ or 15 making 8 congruences in all.

$\gcd(a, 20) = 10$ when $a = 10$. In this case the congruence is solvable when $b = 0$ or 10, making 2 congruences in all.

In all there are 230 congruences which have solutions.

Section 4

1   Following the proof of the Chinese Remainder Theorem, the variables are: $n_1 = 4$, $n_2 = 5$, $n_3 = 7$, $N = 4 \times 5 \times 7 = 140$, $N_1 = 35$, $N_2 = 28$, $N_3 = 20$, $b_1 = 1$, $b_2 = 3$, $b_3 = 0$.

$$
\begin{aligned}
N_1 x_1 \equiv 1 \pmod{n_1} &\iff 35x_1 \equiv 1 \pmod 4 \\
&\iff 3x_1 \equiv 1 \pmod 4, \text{ so } x_1 = 3. \\
N_2 x_2 \equiv 1 \pmod{n_2} &\iff 28x_2 \equiv 1 \pmod 5 \\
&\iff 3x_2 \equiv 1 \pmod 5, \text{ so } x_2 = 2. \\
N_3 x_3 \equiv 1 \pmod{n_3} &\iff 20x_3 \equiv 1 \pmod 7 \\
&\iff -1x_3 \equiv 1 \pmod 7, \text{ so } x_3 = 6.
\end{aligned}
$$

The solution is

$$x_0 \equiv 1 \times 35 \times 3 + 3 \times 28 \times 2 + 0 \times 20 \times 6 \equiv 273 \equiv 133 \pmod{140}.$$

The smallest positive integer solution is 133.

2   (a) We first solve the three congruences individually.

$$
\begin{aligned}
2x \equiv 3 \pmod 5 &\iff x \equiv 4 \pmod 5 \\
3x \equiv 4 \pmod 7 &\iff x \equiv 6 \pmod 7 \\
4x \equiv 9 \pmod{11} &\iff x \equiv 5 \pmod{11}
\end{aligned}
$$

Now,

$$x \equiv 5 \pmod{11} \implies x = 5, 16, 27, \dots,$$

increasing by 11 to a solution of second congruence,

$$x \equiv 6 \pmod 7 \implies x = 27, 104, \dots,$$

increasing by $11 \times 7 = 77$ to a solution of congruence remaining,

and

$$x \equiv 4 \pmod 5 \implies x \equiv 104 \pmod{385}.$$

(b) As the moduli are not pairwise relatively prime a little more care must be taken with this one.

$x \equiv 1 \pmod 4$ and $x \equiv 3 \pmod 6$ have a unique simultaneous solution modulo 12 (the least common multiple of 4 and 6), since $\gcd(4,6) = 2$ divides the difference $3 - 1$. By inspection this solution is $x \equiv 9 \pmod{12}$.

$x \equiv 9 \pmod{12}$ and $x \equiv 3 \pmod 9$ have a unique simultaneous solution modulo 36 (the least common multiple of 12 and 9), since $\gcd(12,9) = 3$ divides the difference $9 - 3$.

$$x \equiv 9 \pmod{12} \implies x = 9, 21, \ldots,$$

and

$$x \equiv 3 \pmod 9 \implies x \equiv 21 \pmod{36}.$$

3 $2310 = 2 \times 3 \times 5 \times 7 \times 11$ and so we require the simultaneous solution of the system:

$$19x \equiv 7 \pmod 2; \quad 19x \equiv 7 \pmod 3; \quad 19x \equiv 7 \pmod 5;$$
$$19x \equiv 7 \pmod 7; \quad 19x \equiv 7 \pmod{11}.$$

Solving these individually gives the equivalent system:

$$x \equiv 1 \pmod 2; \quad x \equiv 1 \pmod 3; \quad x \equiv 3 \pmod 5;$$
$$x \equiv 0 \pmod 7; \quad x \equiv 5 \pmod{11}.$$

Now

$$x \equiv 1 \pmod 2 \implies x = 1, \ldots, \qquad \text{increasing in steps of 2,}$$
$$x \equiv 1 \pmod 3 \implies x = 1, 7, 13, \ldots, \qquad \text{increasing in steps of 6,}$$
$$x \equiv 3 \pmod 5 \implies x = 13, 43, 73, 103, 133, \ldots,$$
$$\text{increasing in steps of 30,}$$
$$x \equiv 0 \pmod 7 \implies x = 133, 343, 553, 763, 973, \ldots,$$
$$\text{increasing in steps of 210.}$$

The order in which the individual congruences are incorporated does not matter. We have usually started with the one with largest modulus, but here we have chosen the opposite order.

Finally,

$$x \equiv 5 \pmod{11} \implies x \equiv 973 \pmod{2310}.$$

4   Translated into a problem in simultaneous congruences we have to solve the system:

$$n \equiv 0 \pmod 9; \quad n \equiv 15 \pmod{16}; \quad n \equiv 23 \pmod{25}.$$

Now

$$n \equiv 23 \pmod{25} \implies n = 23, 48, 73, 98, 123, 148, 173, 198, 223, \ldots,$$
$$\text{increasing in steps of 25,}$$
$$n \equiv 15 \pmod{16} \implies n = 223, 623, 1023, 1423, 1823, 2223, \ldots,$$
$$\text{increasing in steps of } 16 \times 25 = 400.$$

Finally

$$n \equiv 0 \pmod 9 \implies n = 2223 \pmod{3600}.$$

5   The arithmetic progression comprises numbers of the form $11k + 7$, so
    three consecutive terms of the sequence are $11n + 7$, $11n + 18$ and
    $11n + 29$. Our task is therefore to solve, for n, the system of linear
    congruences:

$$11n + 7 \equiv 0 \;(\mathrm{mod}\; 2); \quad 11n + 18 \equiv 0 \;(\mathrm{mod}\; 3); \quad 11n + 29 \equiv 0 \;(\mathrm{mod}\; 5),$$

which, after simplification becomes

$$n \equiv 1 \;(\mathrm{mod}\; 2); \quad n \equiv 0 \;(\mathrm{mod}\; 3); \quad n \equiv 1 \;(\mathrm{mod}\; 5).$$

The first and third of these congruences require that $n \equiv 1 \;(\mathrm{mod}\; 10)$, so

$$n = 1, 11, 21, \ldots .$$

The first number in this sequence to satisfy the remaining congruence,
$n \equiv 0 \;(\mathrm{mod}\; 3)$, is $n = 21$. The three numbers in the arithmetic
sequence are therefore 238, 249 and 260.

6   To be in a position to apply Theorem 4.2 we must first solve each
    individual congruence.

$$5x \equiv c \;(\mathrm{mod}\; 12) \iff 25x \equiv 5c \;(\mathrm{mod}\; 12) \iff x \equiv 5c \;(\mathrm{mod}\; 12)$$
$$7x \equiv 2 \;(\mathrm{mod}\; 30) \iff 7x \equiv -28 \;(\mathrm{mod}\; 30) \iff x \equiv -4 \equiv 26 \;(\mathrm{mod}\; 30)$$

The system has solutions when $\gcd(12, 30)$ divides $26 - 5c$; that is,
when 6 divides $26 - 5c$.

$$5c \equiv 26 \;(\mathrm{mod}\; 6) \iff 5c \equiv 20 \;(\mathrm{mod}\; 6) \iff c \equiv 4 \;(\mathrm{mod}\; 6).$$

So the system has a simultaneous solution when $c \equiv 4 \;(\mathrm{mod}\; 6)$.

7   The number of gold coins n satisfies each of the congruences

$$n \equiv 4 \;(\mathrm{mod}\; 17); \quad n \equiv 10 \;(\mathrm{mod}\; 16); \quad n \equiv 0 \;(\mathrm{mod}\; 15).$$

Now

$$n \equiv 4 \;(\mathrm{mod}\; 17) \implies n = 4, 21, 38, 55, 72, 89, 106, \ldots,$$
$$n \equiv 10 \;(\mathrm{mod}\; 16) \implies n = 106, 378, 650, 922, 1194, 1466, 1738, 2010, \ldots,$$
$$\text{increasing in steps of } 16 \times 17 = 272,$$

and

$$n \equiv 0 \;(\mathrm{mod}\; 15) \implies n \equiv 2010 \;(\mathrm{mod}\; 15 \times 16 \times 17).$$

The least possible number of coins is 2010.

INDEX

$\mathbb{Z}$ 10
$\mathbb{Z}_n$ 10
$\equiv, \not\equiv$ 6
'equivalent to' symbol, $\iff$ 13
'implies' symbol, $\implies$ 32

cancellation rule   13
casting out nines   14
Chinese Remainder Theorem   23
complete set of residues   9
congruence class   9
congruent modulo n 6

divisibility by 11   15
divisibility by 9   15

equivalent congruence   20
Euclid's Lemma for prime divisors   14

incongruent modulo n 6
integral polynomial   16

least absolute residues   9
least positive residues   9
linear congruence   18

number of solutions of a polynomial congruence   17

polynomial congruence   16
properties of congruence   7

residue class   9

simultaneous linear congruences   22
simultaneous solution   23
solution by exhaustion   18